INFORMATION AGES

INFORMATION AGES

Literacy, Numeracy, and the Computer Revolution

MICHAEL E. HOBART AND
ZACHARY S. SCHIFFMAN

THE JOHNS HOPKINS UNIVERSITY PRESS
BALTIMORE AND LONDON

Published 1998
Printed in the United States of America on acid-free paper

9 8 7 6 5 4 3 2 1

The Johns Hopkins University Press
2715 North Charles Street
Baltimore, Maryland 21218-4363
www.press.jhu.edu

Library of Congress Cataloging-in-Publication Data will be found at the end of this book.

A catalog record for this book is available from the British Library.

ISBN 0-8018-5881-X

For our parents, partners, and progeny ■

CONTENTS

ILLUSTRATIONS

ACKNOWLEDGMENTS

The origins of this book trace to a late-night conversation between its authors at the Heartland Cafe, a self-styled "alternative" watering hole situated in funky Rogers Park, a northside Chicago neighborhood. Several years ago, on a balmy summer night, the Heartland's wide beer selection collaborated with the neighborhood's unconventional ambience to inspire (we like to think) a number of rather unconventional ideas concerning science, culture, and information technology. Covering such a broad historical sweep and topical diversity, those ideas would have languished at the Heartland were it not for the subsequent help and encouragement of numerous friends, colleagues, and readers from many disciplines and backgrounds.

We have acknowledged in the Bibliographical Essay following the narrative much of the wide-ranging scholarship on which we have relied. Here, more particularly, we wish to thank several individuals who gave the entire manuscript careful and critical readings from different and contrasting perspectives. With an unfailing instinct for discovering troubled passages, concepts, and arguments, philosopher George Panichas forced clarity upon us—even when we thought we had it—and was a continual source of insight and buoyancy. Historian of science David Lux helped us place materials in proper context throughout the manuscript, adding a wealth of nuance and knowledge to our reading of the interactions between science, technology, and culture. Computer whiz Stuart R. Sheedy combined a retentive memory for detail, an unflagging sensitivity to prose style, and an erudition born of a lifetime of reading and curiosity to give us what every author needs: the critical and informed reaction from a general, educated reader (a dying breed in our culture). The anonymous (to us) historian who read the manuscript for the Johns Hopkins University Press performed an invaluable service by calling to our attention the difficulties of trying to reach both a general and an academic audience. We continue to believe that it can be done; we also believe our attempt to do so has been considerably improved because of the reader's remarks. Also, the editorial staff at the Johns Hopkins University Press, especially Executive Editor Henry Y. K. Tom, has given us sustained encouragement and critical advice.

Other individuals who read and commented critically on several chapters or portions of the manuscript at different phases in its development include D. R. Woolf, June Sochen, Susan E. Rosa, Joan L. Richards, and W. Jay Reedy. Alan Charles Kors, Ann Blair, and Chester Piascik each contributed in a distinctive and timely fashion to resolving specific conceptual, bibliographical, and technical issues. Graphic artist Peter Elsworth prepared the mathematics drawings in a consummately professional manner. Conny Sawyer not only bolstered our entire project with her noteworthy skills as a first-rate secretary and office manager but also supplied an inexhaustible fund of tolerance and good cheer.

Without the active engagement of the above-mentioned individuals this book would have faded like the balmy Heartland moment of its conception. Moreover, in our efforts to capture, develop, and fix in prose that ineffable moment, our readers have saved us from countless errors of fact, judgment, and interpretation. That errors no doubt remain attests our own stubbornness at listening even further to their counsel.

At various stages this project has received institutional assistance from a number of sources. Our home institutions—Bryant College in Smithfield, Rhode Island, and Northeastern Illinois University in Chicago—granted sabbatical leaves to launch the writing, plus some research and travel funds to help underwrite our long-distance collaboration. The National Endowment for the Humanities supported our work with a Summer Research Stipend and with three Summer Seminars for College Teachers, directed by Leonard Barkan, Joan Richards and Shirley Roe, and James Schmidt. Our appreciation extends to all the participants and, especially, to the directors of these fine seminars. In letter and spirit they fully realized the ideals intended by the NEH with its Summer Seminar Program. A Grant-in-Aid from the American Council of Learned Societies and an Andrew W. Mellon Postdoctoral Fellowship in the Humanities at the University of Pennsylvania provided both research funding and time for the project in its early formation. Preparing the final version of the manuscript was assisted by a fellowship from the Oregon Humanities Center, located at the University of Oregon in Eugene. A special word of recognition goes to the center, its director, Steven Shankman, and associate director, Holly Campbell, for creating an atmosphere that exudes the life of the mind at its best.

Finally, in a different category altogether, we wish to acknowledge the long-suffering forbearance and support of our spouses and children: Sally, Abigail, and Julia (for MEH); Amy and Aaron (for ZSS). We have tried our utmost to exhaust their patience, and they are with us still.

INFORMATION AGES

INTRODUCTION

Information Present and Past

Very deep is the well of the past. Should we not call it bottomless?

—THOMAS MANN, *Joseph and His Brothers*

Ours is trumpeted as the "information age," and on the face of it, the claim requires no justification. We are deluged with information, accumulating by the millisecond on video and audio tapes, film, microfiche, floppy disks, hard drives, and memory chips, and spewing incessantly over airwaves, light waves, television cables, and telephone wires. We are surrounded by the "information sciences," propped up by "information theory," crisscrossed (some say unified) by an "information superhighway," spending "information money," performing "information work," fantasizing "silicon dreams."

Yet, like a TV in a waiting room, our electronic, computer culture so mesmerizes us that we fail to recognize other information ages associated with earlier technological changes and to appreciate the historically rooted nature of our own era. Journey back five hundred years to the advent of printing. Crude and mechanical though it may seem to us now, this invention, coupled with the manufacture of cheap paper, caused an information revolution at least as momentous as the one we are currently experiencing. The resultant flood of books swamped the traditional way of thinking based on ancient Greek philosophy, which was ultimately replaced by the modern scientific view of the world, source of the blessings (for some, the curse) of our own age.

Go back another twenty-three hundred years before printing, to the creation of the Greek alphabet, the first form of writing capable of capturing the nuances of speech. Though hardly even an invention by our modern electronic standards, this simple accomplishment constituted a technological revolution arguably more momentous than those spawned by printing or computers. It spurred the Greeks to the kind of speculation upon language and its relation to experience that culminated in the rise of natural philosophy, comprising what we today call "philosophy" and "science." This new form of knowing organized information about the world in a hierarchical system that mirrored the observable order of nature. Greek natural philosophy would remain the intellectual standard for two millennia, providing the bedrock of Western thought and offering a view of the world just as satisfying and at least as commonsensical as modern science.

Drop back yet another twenty-four hundred years before the Greek alphabet, to the invention in Mesopotamia of what was in all likelihood the first form of writing. This technological change, though it jars our modern sensibilities to term it so, was the most momentous of all. For the invention of writing actually gave birth to information itself, engendering the first information revolution. Writing created new entities, mental objects that exist apart from the flow of speech, along with the earliest, systematic attempts to organize this abstract mental world. Here we find the roots of the activity that would ultimately lead the Greeks to correlate the order of the mental world with that of nature. Thus, when we tear ourselves away from the engrossments of electronic culture, we discover that our information age is but the latest of several. From a historical perspective, perhaps the only "information age" truly deserving the title is the original, primeval one of some five thousand years ago.

And this realization plunges us yet deeper into the well of the past. Information came into being at the cusp between orality and literacy, a singular moment that cannot itself be understood unless we first consider the nature of the oral culture that literacy would transform. In this oral world, memory functioned in sharp contrast to the way we literates conceive of it, not as a container for information but as a participatory act, commemoration, serving to maintain social consensus. The emotional power and immediacy of this activity prevented its participants from distinguishing between the content and the experience of commemoration in any consistent manner. Only with this distinction did the mental objects we call information come into being, and only then could memory become a container for them. The information revolution born of literacy is all the more stunning and revolutionary when seen in stark relief against an oral world where information did not exist.

The claim that information once did not exist, that it has a history, sounds absurd. Today's scientists, engineers, and technicians, not to mention social scientists, scholars, and bureaucrats of all stripes, seek and find information everywhere. From the movements of subatomic quarks to the evolution of the entire cosmos, many natural systems are understood as governed by the information they receive and process. Social systems too, in what some pundits tout as the "postindustrial" age, are seen as structured by information, rather than, say, production. (In the 1990s only about a seventh of the American working population actually makes things; the rest receive, process, and transmit information . . . or serve the information moguls.) Information is now commonly depicted as a general principle of organized phenomena, one standing in stark antithesis to the "disorder" or "noise" associated with entropy and the second law of thermodynamics.

Information has become the dominant metaphor of our age, through which we understand ourselves and our world. That we can even envision other historical epochs as information ages attests the metaphor's widespread and growing reach. Yet, when we stop to think about it, we really do not know much about this idiom. Dictionaries assume it, rather than explain it. We move beyond Webster's initial definition (the "communication or reception of knowledge or intelligence") only to find ourselves deeply mired in our own information age and its associated terminology. Information, we read, is "intelligence," "news," "facts," "data," or "the attribute inherent in and communicated by one of two or more alternative sequences or arrangements of something (as nucleotides in DNA or binary digits in a computer program) that produce specific effects." Little help here. Certainly a dictionary speaks to its own time, appealing to intuitively held terms and images that the age already knows and uses, but it does little to take us beyond, to place idiom and meaning in a broader, historical context.

Etymology takes us a step further toward this goal. The term itself traces back to the Latin verb *informare,* which for Romans generally meant 'to shape', 'to form an idea of', or 'to describe'. The verb, in turn, supplied action to the substantive, *forma,* which took varied, cognate meanings that depended mostly on context. The historian Livy used *forma* as a general term for 'character', 'form', 'nature', 'kind', and 'manner'. Horace applied it to a shoelast, Ovid to a mold or stamp for making coins, while the wily Cicero, among other uses, extended it to logic as 'form' or 'species', his rendering of the Greek *eidos kai morphē,* a philosophical expression denoting the essence or form of a thing as distinguished from its matter or content. The practical notion of 'form' as a last, mold, or stamp remained closely tied to its more abstract, logical meaning, which paired content and container. These connotations passed into the earliest English uses of the verb 'inform' ("to give form" or "character" to, or "imbue" with), which date from the fourteenth century, and from which our noun derives.

Behind the late-twentieth-century idiom, then, are the historically grounded notions of information as something informed, shaped by a pattern, and something preserved, set aside from the immediacy of experience. Each notion requires the other. The pattern, the indwelling form, is an abstraction (from the Latin verb *abstrahere,* "to pull," "drag," or "draw away from"), the product of a reflective mental operation that fixes the flux of experience, both ordering and preserving it. This act involves two closely intertwining movements, (1) "drawing away from" experience, such that we are no longer immersed in it and can see it from a critical perspective, and (2) "pulling" or "dragging" something out of it. The twofold movement of abstraction is the sine qua non of information, without which it cannot exist. The mental act implicit in the etymology of the term has become obscured by the contemporary metaphor's imperialistic reach, which has extended beyond the human world into the natural one. Long before information became the stuff of nature, it was the stuff of mind.

If the word originally denoted a human creation, it is the product of an activity that itself has a history. Rather than attempting to find a single, overarching definition of information, applicable across time and culture, we must seek its unique meaning in each age, where technology and culture combine to isolate different kinds of information. We have already implied that the term 'technology' should be broadly construed to include not only the applied sciences, in their various mechanical and electronic guises, but also simpler, less tangible inventions, such as writing and the alphabet. From the Greek *technē*—'art', 'artifice', or 'craft'—the term denotes the practical arts (whether singly or collectively) and the methods that characterize them. Literacy, though we take it much for granted, certainly falls within this definition, and so too does speech, which is as much an acquired skill as reading and writing.

Though both literacy and speech are technologies, only the former is an information technology, as distinguished from a technology of communications. At base, information consists of mental objects separated from the flux of experience, whereas communication may well partake of that flux. Both writing and speech constitute communication, but of the two only writing extracts the sounds of speech from their oral flow by giving them visual representation. In contrast, evanescent speech is part of that very flow, communicating information without necessarily creating or preserving it. (Note that recorded speech can become information by virtue of being removed from its oral context.)

Because information separates mental objects from the flux of experience, it follows that different information technologies can single out different aspects of experience in different ways, generating different kinds of information.

Just as no overarching definition of information serves for all time, so too no uniform, causal relation stands out between the technologies of information and the ages they demarcate. History is universally messier than the tales we can tell with such linear, billiard-ball images of causality. Rather than being an autonomous force that drives the history of information, technology exists in dynamic interplay with culture, shaping and being shaped by it. Thus, in some instances, technology may foster new forms of information, while in others, it is fostered by them.

The complex interactions of technology and culture have produced three distinctive information ages: classical, modern, contemporary. In the classical age we shall explore how the rise of literacy, culminating in its alphabetic form, enabled the classifying potential in natural language to emerge in symbiotic evolution with the technology of writing. The result was a twin birth: of information itself and of the first information age. By the fourth century B.C. the classical world had produced not only numerous and extensive taxonomies (we think of Aristotle here) but also what we might call a classifying mind-set, which assumed that all knowledge worthy of the name could be brought into a properly devised system of general and specific categories. Mastery of the principles of this system brought the mind into harmony with the cosmos, yielding a distinctive information idiom, "wisdom."

In the modern age, commencing with the rise of printing during the Renaissance, typographic literacy did not so much spur the emergence of a new mentality as overwhelm the old one. The surfeit of books and information generated by the print revolution contributed directly to the overburdening and rupture of traditional forms of classification. In turn, this rupture helped clear the way for new, more abstract means of managing information. These means, which we shall designate collectively as "the analytical vision of knowledge," derived from the technology of numeracy, from the newly emerging, symbolic language of mathematics. The modern information age culminated in the eighteenth century with the first of the great modern encyclopedias. Its editors sought to subsume the classificatory tradition under the new, analytical vision, which would produce a reliable, mathematical *mappemonde* or "world map" of reality. The reduction of phenomena to mathematical equations (and the principles they express) underlay a shift in the conceptual foundation of science, as the age of wisdom gave way to that of "knowledge."

Our contemporary information age also has its roots in numeracy. But in the nineteenth century, the mathematical imagination soared into abstract universes far beyond the fantasies of the early theorists who had first devised the idiom. The analytical vision became increasingly attenuated from the material world it purported to map, until the tie finally snapped. Torn from its philosophical foundations, analysis became pure technique, the manipulation of arbitrarily designated symbols according to fixed, logical rules. The pure technique of analysis has fabricated a new home for itself in the electronic circuits of the digital computer, engendering our contemporary information technology and idiom. These, in turn, have fostered a new form of knowing based on the idea of emergence, which describes how certain complex, natural systems continually adapt themselves to their environment. Unlike the analytical vision, this new form of knowing is expansive rather than reductive and open-ended rather than closed. At its heart lie the twin principles of electronic computing, "power" and "play," which define our contemporary information age just as surely as wisdom and knowledge did its predecessors.

In speaking of information 'ages', we shall be using the term not only as a noun, denoting three cultural moments, but also as a verb. In this capacity, ages calls our attention to the movement of information along three different trajectories: reflection, abstraction, displacement. First, it describes the natural tendency of information, once singled out from experience, to become the object of further reflection. This tendency provides the impetus for developments within the classical age, as literacy evolves from its pictographic to its alphabetic form, which more fully realizes the classifying possibilities of language. And this tendency also helps explain the evolution from the modern to the contemporary age, as reflections on the symbolic language of mathematics lead to the hardware and software of the computer. It does not, however, explain the transition from the classical age to the modern one, where the shift from the technology of literacy to that of numeracy marks less an evolution in thought than a rupture.

On another plane, ages places our three cultural moments in a sequence of ever-growing abstraction. Each form of information is separated out from the flux of the world, from the immediacy of experience. And when viewed sequentially, the three ages describe a propensity to draw farther and farther away from the world. Tied to natural language, the classifying impulse of the classical age remains very much rooted in the senses, which provided direct access to reality. The analytical impulse of the modern age is a further step removed from that reality, which became translated into a new language of mathematical symbols corresponding to phenomena. And in our contemporary age, the analytical impulse has become yet farther removed from reality, now rendered digitally as a coded sequence of zeros and ones, themselves without content.

Finally, in its verb form, ages functions in yet another dimension, as new ways of making sense of the world displace old ones. Of course, the old idioms continue to develop. With the dawning of the modern age, for example, science ceased to be grounded in wisdom, but classification entered its golden, Linnaean era; and the electronic computer has given the classifying mind-set a whole new lease on life. Yet, at the same time, classification has been progressively relegated to secondary and tertiary status—we've been there, done that. The process of displacement thus shifts attention from one set of concerns and phenomena to another, as each information age coalesces around its own distinctive set of questions, absorbing and recasting what it can from its predecessors, pushing aside as irrelevant what lies beyond its own cultural ken.

Tripartite divisions of the past, especially into classical, modern, and contemporary ages, are a hallmark of the Western intellectual tradition and provide a broad, albeit Eurocentric, perspective for a history of information. Although we shall touch upon non-European developments—the Mesopotamian origins of writing, the Indian origins of the modern number system, the Islamic origins of algebra—vast portions of the globe and topics, such as the development and influence of writing in China, lie beyond the scope of our project. To approach this universal story from a universal perspective would have rendered it unmanageable. Moreover, the story of information has become universal largely through the diffusion of Western developments, such as the advent of the alphabet, the impact of printing, the rise of mathematics as an instrument for understanding nature, and the invention of the computer. So, although we have left important topics unexplored, our tripartite, Eurocentric perspective is quite appropriate for an initial venture onto uncharted seas.

Like captains of sailing ships, authors of broad-sweeping books must negotiate a number of shoals. Our Scylla looms in the form of overly general remarks, so expansive as to be vacuous, platitudinous, silly. We have tried to avoid such empty comments even as we have had to speak quite generally about a wide range of topics. Only the accumulated scholarship of many specialists in many different disciplines has stiffened our courage to do so. Our Charybdis beckons from the other shore, tempting us to become technical, to speak in the tongues of specialized languages, to clutter our historical narrative with jargon, and to talk thus to fewer and fewer readers. To avoid this shoal we have sought what we may call "rhetorical" explanations in ordinary English, even of technical, symbolic, or highly specialized phenomena. Thus we shall be explaining some of the techniques of Aristotelian philosophy, of mathematics, of symbolic logic, and of computer programming, without assuming any technical or specialized knowledge from the reader.

In the end, to paraphrase the great Swiss historian of culture Jacob Burckhardt, we can only claim to have written an essay in the strictest sense of the word, a trial, an attempt. Our motivation for doing so is quite simple: to employ our training as European historians in the service of a deeper understanding of our own age. Though some specialists may think we have trod their fields cavalierly, we ask them to view their disciplines afresh, from a broader perspective. And although some general readers may balk at exploring these same fields, regarding them as too complex and confusing, we urge them onward. Of both we ask the patience to judge our effort not simply in its parts but as a whole, that the past might serve as a worthy mirror for the present.

PART I

The Classical Age of Literacy

CHAPTER ONE

Orality and the Problem of Memory

"It's a poor sort of memory that only works backwards," the Queen remarked. —LEWIS CARROLL, *Through the Looking Glass*

The Mirror of Literacy

"Sing, goddess, the anger of Peleus's son Achilles and its devastation, which put pains thousandfold upon the Achaeans." Thus, pounding out the rhythm of the meter with his staff, the bard launches into the great song about the baleful consequences of Achilles' unbridled anger. The townspeople, gathered in the marketplace for a festival, know the story well: How King Agamemnon, leader of the Achaean Greeks in their ten-year siege of Troy, captures and enslaves as his concubine the daughter of Chryses, priest of Apollo; how Chryses prays to Apollo for vengeance after Agamemnon haughtily spurns the priest's generous offer of ransom; how Apollo inflicts a plague upon the Greeks, causing the hero Achilles to question publicly Agamemnon's judgment; how Agamemnon, faced with this open challenge to his authority, reluctantly surrenders his captive, only to seize Achilles' concubine in recompense; how the dishonored Achilles sulks in his tent while the Greeks risk defeat in the absence of their greatest warrior. All this the bard deftly portrays in a few dozen measures, setting the stage for heroic struggles to follow.

The townspeople settle in for a special treat. The bard is one of the better of his tribe of rhapsodes, and the song is a long one, suffused with many hours of nonstop action and noble sentiment. Trading his staff for a lyre, the bard sings of the regions and cities of Greece that contributed to the Trojan expedition.

He lulls the people with his rhythms, and their hearts begin beating as one. Their lips move with his, and the murmur of the familiar swells to a shout—"Hurray Ithaca!"—as he recounts their tiny island's contingent of twelve, red-painted ships, led by the wily Odysseus.

Not missing a beat, the bard sings on, of Greek defeats for want of Achilles, of Agamemnon's belated attempt to make amends with the sulking warrior, of Achilles' obstinate refusal to set aside his anger. Measure by bittersweet measure, the song enraptures the listeners, immersing them in the inevitable clash of heroic wills and its bloody outcome. They experience the heat of battle, feeling the ghastly, mortal blows to head, neck, chest, and thigh, seeing with the blood-drained vision of the dying, and reveling in the glory of the victors.

The song builds to its familiar climax, transfixing the townspeople, who hear it as if for the first time. One after another, Greek heroes fall (even their beloved Odysseus lies disabled) until the mighty Hector—son of Priam, king of Troy—seems on the verge of victory. Only then does Achilles relent, ever so slightly, lending his famous armor to his dear friend Patrocles, who dies at Hector's hands after rallying the Greeks. The audience shares Achilles' shock and fury at this loss. He takes to the field and slays Hector, stripping him of bloody armor, piercing his heels, and dragging his torn body to the Greek encampment as food for the dogs.

Now, after many hours of song, the rhapsode brings the story to a close, making his audience feel the unbending of wills. Two heroes have fallen, one Greek, the other Trojan. Achilles reconciles with Agamemnon and holds an elaborate funeral for Patrocles. The townspeople, as if spectators to the funeral games that follow, admire feats of physical prowess and, more so, venerate displays of the better part of honor, as proud warriors vie with each other in proof of the kind of courtesy that Achilles and Agamemnon have demonstrated.

At long last, the bard comes to the final unbending. Achilles takes pity on Priam, who had suffered the worst that can befall a father, watching from the battlements of Troy the merciless slaughter of his son. Tears come to their eyes as the townspeople relive his sorrow, and that of his city. The noble Achilles releases the body, which the gods have protected from defilement, and the Trojans lay their fallen hero to rest, knowing that his death spells their doom.

Reading the epic known to us as the *Iliad* is vastly different from the preliterate experience of hearing and seeing it performed. In place of the bard's galvanic flow of sound and image, the reader beholds a mute tome, the size of a longish novel. It consists of 15,693 lines of dactylic hexameter, divided into twenty-four books, ranging in length from 424 to 909 lines (these figures according to the standard version of the text, established some five hundred years after the song was first sung). The reader can access any one of these fifteen thousand lines at will, going backward as well as forward, reviewing the text to pick up details missed at first glance. (With a book in our hands, we can afford the luxury of inattention.)

Unfamiliar with the term 'dactylic hexameter'? Look it up! For the reader, words are not evanescent sounds but letters on a page, with meanings enshrined in the dictionary. Thus 'dactyl' connotes "a metrical foot of three syllables" and 'hexameter' "a line of verse containing six metrical feet." And these words have etymologies, revealing their historical roots. 'Dactyl' derives from the Greek word for finger, and 'hexameter' from the Greek expression for six measures. By contrast, the preliterate bard would have had no conception of what we term 'dactylic hexameter', though he might sing in the manner of his forefathers about, say, a king with bejeweled fingers, drinking six measures of wine at a feast. For the bard, words have little meaning beyond the concrete things and situations familiar to him and his audience.

Of course, he sings to the audience of its remote ancestors, the Achaeans, and of heroes long dead. But even when he consciously harkens back to olden days, he cannot appeal much beyond the collective memory of the elders in the community, lest he cease making sense. Certain poetic conventions mark the exception to this rule, such as when he has heroes fight with bronze weapons rather than iron, but otherwise he is hemmed in by the present. Compare his situation to ours, where the words here, on this page, can be read in ten or a hundred years as they are now, and in Ithaca, New York, as well as Greece. Writing permits communication over space and time, whereas the bard is constrained by the here-and-now, communicating face to face.

The predominance of face-to-face communication has momentous consequences, both for the nature of oral experience and for what we can infer from it. To his contemporaries, the bard's words are not letters on a page but gusts of air and spittle, inseparable from his gestures and facial expressions. We know from personal experience that the emotional power and immediacy of speech far exceeds that of writing—think only of the difference between a shouting match and an angry letter. And who cannot distinctly recall shouting matches while having long since forgotten what they were about? In the oral world, the face-to-face mode of communication inevitably intrudes upon its content, making it difficult to separate the two, to analyze one as distinct from the other. Though the bard's audience may well recognize and reject departures from the traditional song, it does not listen critically to the bard, as we might study the text of the poem, but participates in the action. Swept up into the rhythmic patterns of word and sound, it inhabits the images they evoke.

We can infer that the audience's difficulty in distinguishing the experience from the content of face-to-face communication discouraged certain kinds of mental operations in the oral world. Recall the sine qua non of information, the twofold movement of abstraction that underlies it, regardless of its form. In the oral world, participants are less likely to "pull" or "drag" something away from the experience of face-to-face communication because they cannot themselves readily "draw away from" the communicative event, viewing it from a critical perspective. The immediacy of face-to-face communication militates against sustaining this twofold movement consistently and systematically. In other words, it militates against distilling information from experience.

Language, of course, continues to function in its ineffable way. One hears and sees and understands. Facts get communicated: "Watch out for that boar!" "The enemy lies over yonder." "It looks like rain." But these facts are not abstracted from the specific circumstances of *this* boar, *that* enemy, *those* clouds, constituting a separable body of knowledge about hunting, tactics, the weather. They do not become information.

This assertion jars us so because the oral world is counterintuitive for us literates, a world apart. We need to enter it in our quest for the origins of information, but before doing so we must abandon our literate intuitions, lest we commit the error of historical anachronism, misinterpreting the past in the light of the present. Some anachronisms are obvious. No English yeoman, kissing wife and child goodbye, ever said he was off to fight the "Hundred Years' War." Other forms of anachronism intrude more subtly, especially those arising from intuitions that, by their very nature, appear universally valid. These can be profoundly misleading, causing us to see in the past only our own reflection. Literacy raises just such a mirror.

We think of the oral world as "preliterate," as characterized by the absence of something we have. Try as we might, we cannot help but see it in the light of this absence. (The alternative—seeing in it the presence of something we have lost—would hardly occur to us.) Thus we are taught in school that poetic song is a form of memory before the invention of writing, a lesson borne out by the so-called discovery of the ruins of Troy in the late nineteenth century. (The site eventually yielded nine "Troys," of which level VIIA may correspond to the city of Homeric legend.) Archaeology seems to have confirmed the power of song as oral memory.

Or has it? Just because an archaeologist claims to have discovered the ruins of Troy does not mean that bards sang the *Iliad* to preserve the memory of the Trojan War. Stated another way, we must entertain the possibility that oral memory functions differently than we literates assume it should. For us, the term 'memory' evokes the image of a thing, a container for information, or the content of that container. Thus, from our literate viewpoint, the *Iliad* preserves knowledge of the Trojan War. But in jumping to this conclusion, we lose sight of the *Iliad* as an oral phenomenon, as the singing of a song. It is not so much a thing as an act, a gestalt uniting bard and audience in a shared consciousness. This phenomenon has little in common with that desiccated thing we literates call "memory." In the world before writing, memory is the social act of remembering. It is commemoration.

The commemorative act looks not so much to the past as the present and future. It provides the community a way of continually redefining itself and its aspirations amid ever-changing circumstances. Commemoration binds the community together as a living entity rather than passively storing information about it. Indeed, the very concreteness of the act, rooted in the here-and-now of face-to-face communication, undercuts the possibility that songs like the *Iliad* systematically preserved information. And if the songs devoted to the maintenance of cultural continuity did not serve as mnemonic containers, information—the stuff we abstract from the flow of experience—could not subsist as something apart from that flow. Our history of information ages therefore begins by examining the nature of memory in oral culture, where commemoration precluded information.

Before substantiating this novel view of memory, we must confess to having employed thus far a sleight of hand, proceeding as if the world of the Homeric bards was a purely oral one, devoid of literacy. A convenience in our foregoing sketch of orality, this assumption is too controversial to serve as the foundation of a sustained argument. The Greek alphabet is widely regarded as having originated sometime in the eighth century—its earliest exemplars date from about 730 B.C.—around the same time as the *Iliad.* Whether the composer of this song knew how to write will forever remain a mystery. Nevertheless, it is by now widely accepted that the bard learned to compose aurally (by ear), listening to predecessors who bequeathed to him a poetic tradition passed on orally (by mouth) from generation to generation. Therefore, regardless of the status of Homer's literacy, his mode of composition offers a window onto the preinformational, oral world.

The Homeric epics originated from songs dating back to the heyday of Mycenaean civilization on mainland Greece, which arose from the stimulus of Minoan culture on Crete and adjacent islands. After the mysterious disappearance of the Minoans around 1400 B.C., marauding Greeks struck out across the Mediterranean, raiding as far west as Sicily and, to the east, all along the coast of

Asia Minor, the northern portion of which harbors the reputed ruins of Troy. The constituent songs of the *Iliad* and the *Odyssey* date from this period of endemic piracy. (One of Odysseus's many sobriquets is "sacker of cities.") The raids eventually subsided with the demise of Mycenaean civilization, which presumably succumbed after 1200 B.C. to the ravages of an even more warlike people, traditionally known as the Dorians.

Although the Mycenaeans had adapted a form of syllabic writing from the Minoans, its unwieldiness suited it only for the most rudimentary purposes, primarily inventorying the contents of a few palace storerooms. And knowledge of it was apparently lost during the so-called Dorian invasions, after which we find no further evidence of syllabic writing on the mainland. So the Greeks maintained cultural continuity by oral means, chiefly in disparate songs about various heroes, at least until the coming of the alphabet. Around this time one or two bards of extraordinary genius wove these diverse songs into the extended epics known to us as the *Iliad* and the *Odyssey,* works conventionally attributed to "Homer," a semilegendary, blind poet from Ionia, the western coast of Asia Minor.

The study of Homer is inseparable from the story of Homeric scholarship, which has given rise to the very subject of orality and its consequences. Although there exist other sources for the study of oral culture, such as the medieval Norse sagas and the modern compositions of nonliterate bards in Europe, Africa, and the Pacific islands, the Homeric epics remain the best known and most thoroughly studied. While steering clear of the intense debates that beset virtually all aspects of Homeric scholarship, we need to examine closely the contributions of two classicists, whose efforts frame the study of orality.

Milman Parry is widely regarded as the father of modern Homeric scholarship for his path breaking work on "formulas," the compositional building blocks Homer inherited from previous generations of oral poets. Extending Parry's literary insights into the realm of intellectual history, Eric A. Havelock has analyzed the cognitive consequences of the "formulaic state of mind" characteristic of oral culture. Although he elicits a convincing general picture of orality from the Homeric mode of composition, Havelock nonetheless betrays the literate assumption that formulas served as storage devices in the world before writing, likening oral memory to a container. In exposing this anachronism, we shall at the same time be constructing the case for an oral world where memory was limited to commemoration, an act that precluded the existence of information.

Homer and the Oral Tradition

From classical antiquity onward people have revered the Homeric epics as the cornerstones of Western literature, the epitome of poetic art, and have deemed the artist responsible for such sublime creations not only a poet but a sage, maybe the greatest of all. Now and then they caught glimmers of other Homers, mean folk poets, or of no Homers at all, only the collective voice of a primitive people. But the epics as we know them were generally regarded as written works until the late 1920s, when Milman Parry single-handedly set the staid, hidebound world of Homeric scholarship on its ear, pioneering a new field known (despite the oxymoron) as "oral literature."

A young American studying classics at the Sorbonne, Parry argued that stock phrases and expressions in the epics were not poetic lapses (even Homer occasionally "nodded") but compositional devices. Though at first he assumed they were literate contrivances, Parry soon became convinced they were the instruments of preliterate poeticizing. To prove this theory, he undertook several expeditions to the rural Balkans, where he recorded the compositional practices of nonliterate, Serbo-Croatian bards. He was accompanied in his travels by a dedicated assistant, Albert Lord, who carried on his work after Parry's tragic death in 1935, at the age of thirty-three (he died instantly when a loaded gun in his luggage accidentally discharged).

Parry's theory stems from the study of Homeric epithets, like "swift-footed Achilles," "gray-eyed Athena," "divine Odysseus," and "glorious Hector." Modern readers typically find the frequent repetition of these expressions monotonous and annoying. Not only do the epithets strike us as clumsy circumlocutions, ill fitting poetic genius, but they often have nothing to do with the sense of the verse. Parry, however, argued that they conform to the metrical requirements of Homeric hexameter, being characterized by the complementary qualities of "economy" and "scope."

By "economy" he meant that for each kind of metrical situation the poet encountered, there was one — and generally only one — appropriate noun-epithet phrase. Depending upon where the name appears in the meter, "Achilles" must be accompanied by "swift-footed," "godlike," "son of Peleus," or some other epithet, which, in conjunction with the name, meets the requirements of the particular metrical situation. Accordingly, one might find Achilles referred to as "swift-footed" in a line that has nothing to do with running.

The "scope" of these noun-epithet phrases complements their economy. They vary so as to encompass, in the case of each name, virtually all the commonly occurring metrical situations. Parry concluded that the economy and scope of the epithets resulted from a long process of evolution, in which just the right ones were "selected out," creating a traditional stock of phrases.

Parry termed these phrases "formulas," defining a formula as a group of words "regularly employed under the same metrical conditions to express a given essential idea." In addition to the epithets for proper names, he also identified other formulaic expressions in the Homeric poems, such as "wine-dark seas" and "black-hulled ships." And he eventually extended his definition to include not only noun-epithets but also other, larger word groupings—such as conjunction-verb phrases and even acoustically similar phrases—used repeatedly under the same metrical conditions. Some claim that his definition of formulas ultimately became too extravagant, for the longer the formula, the greater the difficulty in determining exactly what elements are being repeated. But this criticism need not detract from his path breaking insight into the formulaic nature of Homeric composition, a lost art whose richness we literates can barely comprehend.

The epics utilize many patterns of composition (involving meter, rhythm, phrase, verse) whose study is quite technical. We shall mention only one other stock element, the "theme," elaborated by Albert Lord. Themes usually depict events, such as assemblies, journeys, and battles. These, in turn, consist of subthemes, such as the calling of an assembly, the preparation for a journey, and the arming for a battle. Echoing Parry, Lord defines themes and subthemes as "groups of ideas regularly used in telling a tale in the formulaic style of traditional song." Their essential characteristic is repetition. For example, when different heroes arm for battle, they all tend to do the same things in the same order.

Parry and Lord pioneered the study of oral literature, showing how formulas and themes enabled nonliterate Serbo-Croatian bards to compose in performance works similar in length to the Homeric epics. Needless to say, if these bards had to weigh the metrical qualities of every word and phrase, and chart every move in the narrative, composition would have been a lengthy process that could only occur before performance. But because they commanded a traditional storehouse of expressions and events, refined over generations of usage to fit prescribed poetic and narrative requirements, composition could occur concurrently with performance. Armed with the evidence of Balkan practices, Parry and Lord contended that our literate versions of the *Iliad* and the *Odyssey* originated as oral works.

Students in the field established by Parry and Lord have subsequently revealed other models of oral literature, neither exclusively formulaic nor composed solely in performance. Modern nonliterate bards in Africa and the Pacific islands, for example, memorize all or part of their compositions before delivering them. But formulas and themes obviously enhance the process of oral delivery, helping bards ancient and modern either to deliver long compositions from memory or to compose them in performance. Formulas and themes also enable audiences to remember these bardic productions and to criticize deviations from accepted metrical and narrative patterns. Of course, some change inevitably occurs from one performance to the next, but the formulaic tradition helps preserve oral literature relatively intact from generation to generation.

Enter the influential yet controversial classicist Eric A. Havelock. The revelations of Parry and Lord posed for him an obvious question: Why did the formulaic tradition arise in the first place? In his provocative *Preface to Plato* (still in print after more than thirty years), he shocked many of his colleagues by asserting that poetry in oral culture serves not aesthetic but practical purposes, the preservation and transmission of information necessary for the survival of society. Taken together, the *Iliad* and the *Odyssey* constitute for him a Homeric "encyclopedia," in the modern sense of the term as a compendium of knowledge and in the original Greek sense of a "circle of learning." According to Havelock, the epics circumscribe all one needed to know in the oral culture. Repeated exposure to this body of knowledge through the performance of the poems constituted the chief process of education in pre-Socratic Greece.

Everywhere he looked in Homer, Havelock saw a wealth of instruction. For instance, the quarrel between Achilles and Agamemnon at the beginning of the *Iliad* embodies for him a wide range of subliminal "teachings." It lays out the rules for the disposition of captives, the etiquette of making and receiving ransom requests, the reverence due to priests, the respect accorded kings by powerful warriors, and the symbols of public authority and their functioning in a warrior assembly. In addition to these "social," "political," and "religious" lessons, Havelock also identifies practical ones, such as instructions on how to launch and land a ship.

The epics preserve this information in what Havelock calls "formulaic passages," identifiable by virtue of their repetition. Whenever a hero addresses an assembly, sacrifices to the gods, or embarks on a ship, he follows in each case the same series of actions, performed in the same ritual order. Although instances of the same action follow in the same order, they are expressed in different passages of the poem by different "verbal formulas," which Havelock (like Parry) defines as "those building blocks made up of rhythmic units of two or more words recurring in identical order and in identical place in the line." The wide range of verbal formulas allows the bard to teach the same lessons over and over again without becoming boring.

For Havelock verbal formulas simply function as the means of expressing the lessons concealed within the formulaic passages: "The real and essential 'formula' in orally preserved speech consists of a total 'situation' in the poet's mind. It is made up of a series of standardized images which follow each other in his memory in a fixed order. The verbal formulas serve as the instrument by which these images are deployed." Reduced to a standardized image or situation in the memory of both poet and audience, the values, beliefs, and practices of the oral culture stand behind the formulaic tradition.

From this initial premise, Havelock proceeds to draw some startling conclusions about the cognitive effects of orality, conclusions that, while remaining controversial, provide one of the most extensive and penetrating analyses of oral culture. The oral technology for storing and communicating information created what he terms a "formulaic state of mind," characterized by an active and unreflective participation in the mnemonic process. Havelock's bard sought to sweep the audience up into the rhythm of the song. Caught in his hypnotic spell, the members of the audience identified uncritically with the action in scene after scene. They *became* proud Agamemnon, feeling his kingly refusal to surrender his rightful prize of war, and then they *became* wrathful Achilles, challenging the king's disastrous decision. This process of identification fixed the story uncritically in their minds. Although they might have recognized and rejected deviations from the traditional tale, they could not criticize its subliminal teachings.

These lessons are "storied," embedded in scenes of action and adventure. Rather than enumerating political, religious, and social precepts, book 1 of the *Iliad* tells the story of Agamemnon's defense of royal prerogatives, of the priest's invocation of divine vengeance, and of Achilles' grievances before the warrior assembly. Concrete nouns and active verbs move the story along at a breathless pace. According to Havelock, the bard scarcely employs copulas with the Greek equivalent of the verb "to be," constructions belonging to the language of abstraction. For example, the statement "All warriors are proud" denotes a quality abstracted—in the Latin sense of "drawn from"—a multitude of individuals and their actions. The bard does not utilize the verb "to be" because its constructions cannot be storied; they cannot be incorporated into the epic, into the culture's form of "preserved communication."

If the bard can utilize only narrativized statements, then, according to Havelock, he can think only concretely and not abstractly. And if he cannot think in abstractions, neither can his audience. One might object that the paucity of verbal copulas in the epics does not necessarily imply an audience incapable of using them in everyday speech. But Havelock would insist that the epics constitute the accumulated wisdom of the culture, beyond which the audience (thoroughly inculcated with the teachings of the epics) cannot go. If abstractions cannot be expressed in the epics, they cannot be thought by the audience. And if abstractions are beyond conception, "thinking" as we know it cannot take place.

For us, thinking entails a critical distance between the mind and the object of thought—we think "about" something. The participatory nature of the epics prevented the attainment of that critical perspective. According to Havelock, critical thinking became possible only with alphabetic literacy, which obviated the need for oral mnemonics. With its vowel and consonant signs, the Greek alphabet was sufficiently flexible to translate the sounds of speech directly into writing, thus preserving the Homeric epics in their entirety, without distortion, abridgement, or simplification. Readers of this oral tradition no longer needed to identify with the action of an epic in order to remember it. Writing freed their psychic energy to flow in new directions, toward "a review and rearrangement of what had now been written down, and of what could be seen as an object and not just heard and felt."

Alphabetic literacy fostered a new state of mind heralded by Platonic philosophy, which is characterized by a "separation of the knower from the known." Liberated from the spell of the Homeric chant, the reader could distinguish between his "thinking self" and the oral performances that, previously, the audience had participated in uncritically. "Knowledge" now became possible, as information no longer needed to be embedded in an epic narrative with which one identified emotionally; it could take on "objective" existence outside one's experience.

To the extent that things known became objects, they came to be classed with other, similar objects, leading one to inquire about the essence underlying these similarities. "And so," concludes Havelock, "the [pages of Plato's *Republic*] are filled with the demand that we concentrate not on the things of the city but on the city itself, not on a just or unjust act but on justice itself, not on noble actions but on nobility, not on the beds and tables of the heroes but on the idea of bed *per se.*" The process of abstraction resulting from alphabetic literacy thus eventually found expression in Plato's famous "theory of forms," the basis and epitome of his philosophy.

Some fault Havelock for insisting too strenuously on the pure orality of Homer's world and the widespread literacy of Plato's. The former assertion cannot be proven, and the latter is probably wrong. But these flaws do not go to the heart of Havelock's argument, which concerns less the extent than the effects of orality and literacy. The first to treat these effects at length, his account remains the most detailed and compelling evocation of oral culture.

Havelock's Homer embodies fully the face-to-face mode of communication. On the most obvious level, the bard sings to an audience. On a deeper level, the songs convey scenes of face-to-face communication—when Agamemnon rebuffs the priest, when the priest calls upon Apollo for vengeance, when Achilles challenges Agamemnon's actions. Finally, on the deepest level, the audience, captured by the spell of the song, actually participates in these interactions, becoming the haughty Agamemnon, the vengeful priest, the wrathful Achilles. Each song thus discloses a series of face-to-face communications, nested one within another like Russian dolls. Thoroughly permeated with the here-and-now, these encounters preclude the separation of knower from known.

Although this aspect of his view offers a convincing general picture of orality, Havelock's interpretation as a whole suffers from the anachronistic assumption that oral cultures must have had some means of storing information. In other words, he projects the literate notion of memory as an information container onto the oral world. This anachronism appears strikingly in the imagery he uses to describe the Homeric encyclopedia: "We shall deliberately adopt the hypothesis that the tale itself is a kind of literary portmanteau which is to contain a collection of assorted usages, conventions, prescriptions, and procedures." The quaint phrase "literary portmanteau" betrays Havelock's assumption that memory in an oral culture serves the same function as it does in a literate one, to contain information.

By viewing oral memory in the mirror of literate preconceptions, Havelock mistakenly sees an intellectual divide between oral and literate modes of information storage. He maintains that the participatory aspect of the epics hinders all abstract intellectual processes. By contrast, we shall show how the epics embody certain kinds of abstractions, the kind that aid in commemoration rather than information storage. The real divide separates an oral world where information does not exist from a literate world where it does.

The Nature of Commemoration

Greek oral culture must have had some means of passing on its knowledge, values, and beliefs, but these did not necessarily constitute "preserved communication," a body of information retained in narrative form. In a purely oral culture, knowledge, values, and beliefs exist not as information but as practices whose preservation is a by-product of repeated usage.

Imagine the early polis world of the ninth or eighth century B.C., a world of small, scattered city-states, divided by mountains and existing in limited contact with each other. In such communities, one learns how to behave in the assembly by watching one's elders, an assembly being little more than a group of warriors meeting in the marketplace. Likewise, one learns how to sacrifice to the gods by watching sacrifices. And one learns how to launch and land a ship by watching and participating in these activities. In small, traditional communities, one engages in education everywhere. There is simply no need for preserved communication to store knowledge of everyday practices.

Of course, the epics surely taught something. They helped inculcate the values of the warrior elite. And they may also have provided instruction in practical matters. Book 23 of the *Iliad,* for example, contains a striking passage in which the wise old Nestor advises his young son on how to win a chariot race despite having a slower team of horses. (His advice: Hug the post. Note that Homer undercuts the generality of this rule by giving a detailed description of the specific post in question.) It is hard to imagine that youths in the audience did not learn something about horsemanship from this passage. Moreover, the inclusion of certain political and religious practices in the epics no doubt gave them a special sacral quality that may have helped preserve them relatively unchanged for long periods. But the essential point remains: Practices prevalent in the community did not need to be preserved in the epics because they were everywhere in use.

And when they passed out of use, they were forgotten. In traditional communities change usually occurs so gradually as to go unnoticed. Let us imagine, though, a hypothetical example of sudden change: A sailor discovers how to tie a new kind of knot clearly superior to the old one. The new practice would quickly spread throughout the community. For awhile some might remember the old practice, not as belonging to a body of information about how to tie knots but as an anomaly soon forgotten. Eventually, the new way of tying knots would cease to be new at all, becoming instead the way of one's forefathers, in use from time immemorial. And the old way would cease being "old," becoming instead "wrong." When stored as practice, knowledge of knot tying has only ephemeral existence apart from the activity of sailing that preserves it. (The alternative way of preserving such knowledge in an oral culture would be to maintain a class of professional knot-tiers, hardly an effective use of manpower.)

The assumption that traditional oral communities needed a form of preserved communication reflects literate preconceptions about memory. We moderns regard memory as a container filled with information, a notion strongly reinforced by the terminology of our computer culture, with its "hard drives," "RAM," and "databases." This notion, however, originated long before computers, with the spread of literacy, for writing enables us to convey the same information, with the same truth value, to different people in different times and places. On this account, it fosters what we might call a "textual" model of memory, whereby the mental objects contained within our heads are akin to the pieces of information stored in writing.

The textual model misleads because it emphasizes one function of memory, the storage of information, at the expense of another function, the act of recollection. It focuses on "knowledge" to the exclusion of "remembering." The limits of the textual model quickly become apparent when we consider the actual experience of remembering, which proceeds not only by means of words but also of emotions and sensations, the latter including taste, touch, smell, sound, and sight. Witness the memories evoked for Proust by the experience of seeing, smelling, and tasting a madeleine. Indeed, sensual spurs to memory are much more powerful and prevalent than verbal ones. By emphasizing knowledge to the exclusion of remembering, the textual model obscures the nature of memory as a multifarious activity.

The manifold, sensory nature of the individual act of recollection also characterizes the "collective" or "social" act. Far more than simply the stored knowledge of an oral culture, "social memory" is the act of commemorating, which proceeds not simply by means of words but, more so, by means of the visual images and acoustic patterns that the words preserve and evoke—"wide-ruling Agamemnon," "swift-footed Achilles," "Hector of the brazen helmet." These formulas comprise larger images and patterns that serve as aids in remembering and, thereby, constitute oral forms of abstraction, generalizations distilled from specifics.

All cultures, whether oral or literate, utilize such mnemonic aids, especially the visual ones, which anthropologists have termed 'maps'. "A 'map' is a visual concept, a constructed or projected image, referring to and bearing information about something outside itself," write James Fentress and Chris Wickham in their cogent study *Social Memory.* The word originates from the Latin *mappa,* for "napkin" or "cloth," from which the Middle Ages derived its term *mappa mundi,* or "map of the world"—hence our modern geographical usage. As an aid in remembering, however, a map might best be conceived of as a picture or diagram sketched on a napkin, as if visually illustrating a point in a dinner conversation.

The sketch could represent the floor plan of a house, an electronic circuit, the atoms of a molecule, the plot of a story, or a philosophical concept—anything real or imaginary subject to visual representation can be mapped. We can, for instance, remember a face by mapping its essential features, as in a caricature, or remember a joke by visualizing the types of characters involved, types (doctor, priest, rabbi) that reduce to their essential features. In contrast to the textual model of memory, maps evoke understanding by means of generalized visual images, abstractions not automatically reducible to words.

Whereas maps designate small-scale mnemonic aids, each presenting a single visual image, epics represent large-scale ones, linking together many visual images. Each epic consists of a sequence of scenes or situations that serve to map the action of the narrative. These scenes are linked together by aural cues—spoken words and phrases that are remembered like visual images because they form certain types of patterns. Memory in an oral culture thus involves the recollection of abstracted patterns, both visual and aural, not of words.

Havelock himself supports the notion that epics consist of a sequence of visual images. Recall that he describes the Homeric poems as consisting of "formulaic passages" depicting the same kinds of actions over and over again, and that "verbal formulas" underlie these passages, enabling the bard to convey the same information in infinite variation: "The real and essential 'formula' in *orally preserved speech* consists of a total 'situation' in the poet's mind. It is made up of a series of standardized images which follow each other in his memory in a fixed order" (emphasis added). The images are "standardized" by virtue of being reduced to their essential features, and, as such, they are maps.

Note, however, that Havelock refers to the epics as "orally preserved speech." In keeping with the textual model of memory, he assumes that images merely preserve language, the vehicle conveying information. In fact, the Greek alphabet towers so importantly for Havelock precisely because it can transpose every nuance of spoken language into writing; it offers the most efficient representation of "orally preserved speech." Yet, contrary to Havelock, the images in the epics exist not to preserve language, as in the textual model of memory, but rather to facilitate the activity of commemoration.

In this activity, images become standardized because the community can remember them only by abstracting their general features. Again Fentress and Wickham: "Images can be transmitted socially only if they are conventionalized and simplified: conventionalized, because the image has to be meaningful for an entire group; simplified, because in order to be generally meaningful and capable of transmission, the complexity of the image must be reduced as far as possible." Thus an epithet like "proud Achilles" conjures the general image of a hero—erect, defiant, confident—for members of an audience whose individual notions of pride derive from observing the particular bearing of actual heroes in specific circumstances—in the assembly, in battle, in gymnastics. Commemoration, then, proceeds by reducing such diverse images to a socially defined common denominator.

When we view the epics in the light of this commemorative activity, we see a process of abstraction at work that transcends even the concrete settings, events, and language of poetic narrative. Each epic consists of a series of visual images linked together by words. General enough to be remembered by all segments of the community, these images form the material on which the bard embroiders. As he sings the song, he embellishes each visual image with specific details about personalities, practices, and customs, all of which may vary slightly from one performance to the next, reflecting the changing circumstances of the community. The general image, however, will stay unchanged. It comprises a visual abstraction, distilled from and encompassing a range of specific images.

An acoustic as well as visual process of abstraction preserves the narrative. Again, epics are sequences of images linked together by words. In an oral culture with a formulaic style of composition, both bard and audience remember these words more as sounds than as words per se. From this perspective, formulas might best be understood, not in Parry's terms as "groups of words" fitting certain metrical requirements—a highly literate notion—but rather as patterns of sound abstracted from speech. The immediacy of face-to-face communication, therefore, does not preclude the possibility of abstract mental operations. To the contrary, these comprise the very stuff of social memory.

Why should the community bother to remember such images? Let us begin to answer this question by considering the function of memory in our own individual lives. It certainly preserves facts, but unreliably and inconsistently. Not only do we forget facts about the world around us—Who was the last unsuccessful vice-presidential candidate?—but we even forget the facts of our own personal histories, for which we are the chief repository. We tend to regard this kind of forgetfulness as accidental or unintentional, as reflected in the expression that something has "slipped" our mind. Far from a defect, however, slippage forms an integral part of individual memory, which is a highly selective activity.

By means of slippage we continually reinterpret our lives in the light of current circumstances, downplaying or forgetting events in our personal histories that may once have seemed important. Ongoing and hence gradual, this process makes our perception of ourselves seem stable from one day to the next even though it fluctuates constantly. Our memory of ourselves comprises not so much the collected facts of our personal history as the act of retelling, and thereby reorienting, ourselves in an ever-changing present. Memory as a selective activity provides us with existential stability, not information.

Similarly, social memory designates a selective activity that provides consensus within the community. In his recent study of social memory, *How Societies Remember,* Paul Connerton writes: "Concerning social memory in particular, we may note that images of the past commonly legitimate a present social order. It is an implicit rule that participants in any social order must presuppose a shared memory. To the extent that their memories of a society's past diverge, to that extent its members can share neither experiences nor assumptions." Thus those who came of age in the 1960s think of those who came of age in the 1990s as "Generation X," as being characterized by an unknown factor. Although they both use the same language in everyday speech, these generations supposedly talk past each other because their words refer to different things, different experiences, different texts.

This state of affairs would be inconceivable in an oral culture, where commemoration establishes a common memory of the past, providing all members of the community with the same point of reference. As the true instrument of consensus, social memory does not embody an authorized version of the past, fixed for all time. Instead, the activity of commemoration continually reinterprets the past in the light of an ever-changing present. In so doing, commemoration enables the community both to cohere in the present and to (re)define its aspirations for the future: memory working forward, the White Queen might have said.

Commemoration accomplishes this social reorientation by revising the facts about the past. In an epic narrative, these facts reside in the details of each scene or situation, details concerning, say, the specific nature of social practices. Although such details may vary in the light of ever-changing circumstances, the visual image underlying the scene possesses sufficient generality so as to remain unchanged. In other words, the facts of the narrative fluctuate while the abstract images preserving it remain relatively fixed. As Fentress and Wickham observe, "Social memory is not stable as information; it is stable, rather, at the level of shared meanings and remembered images." By sharing the recollection of the same images, the community remains a community.

From Commemoration to Information

A fundamental difference exists between the oral process of abstraction and literate ones, namely that the oral process is participatory and unreflective. By the latter adjective we do not mean to imply that oral cultures are somehow "primitive," with nothing to teach us literates. Indeed, if the oral interpretation of Homer has accomplished anything, it has shown that Western literature evolved from sophisticated compositional practices whose art is long lost. And formulas represent only the tip of this aesthetic iceberg, the full extent of which we literates can scarcely imagine. Far from implying our own cultural superiority, we term the oral mode of abstraction 'unreflective' in the highly specific sense that it does not foster a critical distance between knower and known.

An unconscious and universal tendency to map patterns, whether visual or acoustic, produces oral abstractions. In an oral culture, these patterns are not perceived as such. The community does not decide in advance, as it were, on the nature of the patterns it will store. Rather, it distills them from diverse memories of an epic through an automatic social process, a reduction to a common denominator. Similarly, although bards compose by manipulating visual and acoustic patterns, they do not have in their heads mental tables of sounds used to stitch together preconceived visual images. Rather, they compose their songs from particular sounds and images that simply "feel right" together—hence, they are "poets."

How do the reflective mental processes that separate knower from known come about? Havelock attributes this development specifically to Greek alphabetic literacy, which translates what is known (the Homeric encyclopedia) into a form that no longer needs to be experienced emotionally but can be examined critically. In this specific sense, what is known becomes separated from the psychic state of the knower. Of all the forms of ancient writing, according to Havelock, only the Greek alphabet faithfully translated a highly nuanced form of "orally preserved speech" into a mental object.

But, contrary to Havelock, this kind of reflective mental process does not hinge on the faithful translation of oral into visual information. Although writing preserves information, it only does so by first creating it. This is the real separation of the knower from the known, which does not conjure abstract thought ex nihilo but simply makes one aware of the abstractions one had previously participated in.

Translated into writing, the participatory images of the epics—experienced by the oral culture as the flow of speech—are "taken out of" the flow and, by that literal act of abstraction, given "form." Recall the etymology of 'information', which traces back to the Horatian and Ovidian sense of 'form' as a shoelast or a mold, imparting spatial organization to matter. This spatial sense of the term implies the Ciceronian meaning, that of differentiating form by "species" or "kind." Thus, when writing informs the oral flow, its contents can be identified, removed from the narrative, and reorganized by type.

All literacy, whether pictographic, syllabic, or alphabetic, separates knower from known. Writing need not faithfully represent speech in order to isolate and shape experience. Indeed, the separation of the knower from the known begins with the very earliest form of writing, which long predates the alphabet and Platonic philosophy. The "Platonic state of mind" attributed to the alphabet is simply the natural outcome of a process of abstraction that originated some five thousand years ago, with the invention of writing in Mesopotamia. That primeval informing of experience constitutes the onset of the first "information age," marking the birth of information itself.

This bald claim may still seem too counterintuitive for us literates. How could something so commonplace as information not exist in an oral culture? Recall, though, that everyday speech readily communicates facts in the oral world without storing them as information. (We literates can think of speech as communicating information only because many different kinds of texts subtend the act of communication, giving stability and order to the facts we cite.) And also recall that technical knowledge in the oral world subsists not as information but as practice. Such knowledge is not informed because it lacks sustained existence apart from the practices that preserve it.

But surely, one might argue, the Greeks knew all sorts of geographical and historical information—that, say, the region of Thrace lay so many days' sail up the coast and that the Mycenaeans were their ancestors. To the extent that everyday speech conveyed this kind of knowledge, however, it remained ephemeral and did not constitute a stable, isolable body of information. The Greeks knew of Thrace because they sailed there; but were they to have stopped, they would have soon forgotten its existence. What we call geographical information resided in the practice of travel. And what we call historical information had an existence even more transitory. Living amid the architectural ruins of their Mycenaean predecessors, the Greeks eventually attributed such massive buildings and walls to a race of Cyclopes.

But surely, one might still object, when ritualized forms of communication store such knowledge, it becomes information; hence, traditional stories about the age of Cyclopes and the Trojan War preserved and passed on historical and cultural data from generation to generation. Here too, though, we remain largely in the realm of practice, the singing of songs being little different from any other traditional activity. To the extent that the epics did convey a cultural heritage, it was simply the by-product of, and inseparable from, the celebration of the ongoing existence of the community. The datum that Greeks had once made war against Trojans was submerged in the oral mind by participatory images serving as the instruments of commemoration: haughty Agamemnon, wrathful Achilles, glorious Hector. And the energy spent on animating these images, on entering into them, precluded drawing away from the narrative sufficiently to extract information from it. The cultural heritage was "hard-wired," as it were, by the process of commemoration, rather than existing (as it does for us literates) as a separable body of information.

But surely, one might insist, the participatory images themselves must constitute information, the stuff of visual maps, and the tendency to map information must have been equally apparent in other cultural products, like

genealogical tattoos and ceremonial drawings. Again, though, we should take heed not to project ourselves as the measure of all things. What we literates might perceive in these maps as information was not experienced as such by the members of the oral culture. For them maps evoked rather than informed. Even though the maps functioned as visual abstractions, the members of the community could not separate them from the participatory images and rituals in which they inhered. And if the maps could not be abstracted beyond commemorative practices, in and of themselves they did not constitute information as a separable body of mental objects.

Information is thus wedded to writing insofar as writing gives stability to the mental objects abstracted from the flow of experience, such that one can access them readily and repeatedly. This stability is inherent in the twofold movement of abstraction that undergirds information. In order to take a mental object out of the flow of experience, one must first draw away from that experience, seeing it from a critical distance that fixes its aspects. When the mind then extracts a mental object, it captures the form or pattern underlying the appearance of things. Reflected in verbs like "capturing," "seizing," "grasping," "apprehending," and "comprehending," the mental activity restrains the flow and hence stabilizes the products of experience. By giving mental objects a sustained existence apart from the flux of the oral world—apart from evanescent speech, apart from practice, apart from ritualized communication and its maps—writing gives these objects a stability they cannot otherwise have. It creates information.

We shall now attend the birth of information in ancient Mesopotamia, when literacy first changed a form of preserved communication that was primarily commemorative into one that was primarily informational. We shall see how this transformation derives from the classificatory potential inherent in speech, how writing made the Mesopotamians conscious of this potential, and how, when they became conscious of it, classification took on a life of its own, manifesting itself in the activity of exhaustive list making. We shall leave the Greeks, our exemplar of oral culture, far behind as we plumb the depths of an even more remote past in search of the origins of information. But we shall return to Hellas again in the third chapter, when the alphabet enables its inhabitants to clarify the principles of classification first glimpsed by the Mesopotamians.

CHAPTER TWO

Early Literacy and List Making

At the basis of all writing stands the picture.
—I. J. GELB, *A Study of Writing*

The Mysterious Invention

Every literate culture seems to know who invented writing. For the ancient Egyptians, it was Thoth, record keeper of the gods; for the Assyrians, Nabu, god of wisdom; for the Greeks, the Phoenician prince Cadmus, founder of Thebes; for the Mayas, Itsamna, son of the Creator god; for the Arabs, at least according to one tradition, the Prophet Muhammad. In addition to these human and superhuman agents, we find nonhuman ones as well. The Ekoi of Nigeria credit monkeys, and one Chinese legend honors the efforts of a sacred turtle. So much certainty about the inventors of writing, but about the nature of the invention, silence.

Alas, its inventors did not have the foresight to claim the credit themselves, at least not in writing, leaving that task to subsequent, literate generations, for whom the nature of the invention was already self-evident. What Egyptian scribe, using hieroglyphs shaped like a vase, bird, snake, leg, or eye, could doubt that Thoth derived writing from picture drawing? And what ancient Greek, whose letters obviously stood for sounds, could doubt that Cadmus invented or (traditions vary) adapted symbols for representing speech? But whether princes, prophets, or primates, the inventors of writing were necessarily preliterate, and thus the nature of the invention could not have been self-evident to them.

It seems a curtain descends whenever we try to cast back, beyond the invention, to its preliterate origins, for literacy assumes itself. Witness our modern definitions of 'writing', which reveal even less than those of 'information'. The dictionary renders the term as the commonplace act or its products that we all take for granted. Even etymology, the trusty fallback of the intellectual historian, offers no help. From the Old English *writan*—itself related to the Old Frisian *writa*, the Old Norse *rita*, the Old Saxon *wrîtan*, and the Old High German *rizan*—the original term and its cognates denote scoring, tearing, cutting, or . . . writing. Try as many languages as we like, the term defines itself.

Scholarship appears incapable of lifting this curtain. In rare unanimity, historians of writing designate it a species of picture drawing, be it pictures of things or, as Voltaire described with characteristic elegance, "the picture of the voice." But at what point does drawing become writing? Most mark the transition with the advent of "phonetization," when pictures of objects come to stand for sounds of speech, yielding "true" writing. By this standard, though, its "truest" form is our own alphabetic writing, a conclusion suggesting the anachronistic character of modern, no less than ancient, interpretations. Is there a way to raise the curtain, to reveal the innovation underlying the invention?

We can start from the consensus that writing grew out of drawing, but we need to find a new standard for essaying afresh the point at which drawing became writing. Thus far we have shown that information did not exist in the oral world, and we have asserted that it comes into being with the advent of writing. Because information identifies so closely with writing, perhaps we can discern qualities inherent in information that distinguish picture drawing from writing. This approach has the advantage of circumventing the self-referential nature of the term 'writing' in the search for the origins of literacy.

Recall that the etymology of 'information' traces back to *forma*, which presumed for Horace and Ovid the act of giving shape to something and for Cicero the related act of sorting shapes by "species" or "kind." Accordingly, to "inform" something, to abstract or take its form out of the flow of experience, implies classifying it. One of the basic functions of language, classification utilizes nouns (from the Latin *nomen*, or "name") and adjectives (from the Latin verb *adjicere*, "to add to"). When we communicate an idea by "adding to a name," we are classifying.

In this regard natural language functions in a fundamentally different way from picture drawing, a holistic form of communication. The picture of a red block is a single image, whereas its linguistic representation (oral or written) combines the name ('block') and what is added to it ('red'), classifying the object by shape and color. The same holds true for a drawing of two blocks versus the linguistic expression, "two blocks." (In fact, long before the advent of arithmetic, numbers functioned as adjectives.)

Writing originated as the need for increasingly complex record keeping developed the classificatory potential inherent in language. Until the need arose, this potential had remained unrealized in the oral world, where classes existed in speech only ephemerally, fleetingly, as part of the practice of everyday life. A shepherd might speak in general of "livestock," or in particular of "sheep," "goats," or "cattle," only to have the class and its subclasses vanish from his mind along with the sound of the words. As long as his remained a small flock, with all its members grazing under his watchful eye, he had no need of classification beyond linguistic convenience.

But when he found himself having to keep track of many different kinds of livestock—hundreds of sheep in field X, hundreds of goats in field Y, and hundreds of cattle en route to the market—he had greater need of the classifications of language to facilitate record keeping. Writing emerged as the natural tendency for classifying found its visual expression and thus reified something once fleeting. The real innovation here, the genius of the invention, lay in abstracting language's ability to classify from speech and giving it a form independent of picture drawing.

This perspective on the origins of writing exposes yet another, tenacious anachronism, no less misleading than the one disclosed in the previous chapter. We generally think about writing (when we bother to think about it at all) as a vehicle of expression and, therefore, as a technology of communication. Writing either stands for the act by which we communicate (say, putting pen to paper) or the product of that act (words on a page). In either case, writing functions as a technology of communication by rendering speech visible. When Voltaire called it "the picture of the voice," he merely expressed the commonplace association of writing with speech.

Yet writing did not spring forth fully formed as a technology of communication, much less one communicating speech. Its genius resides in the fact that it originated as something apart from both picture drawing *and* the spoken word, something absolutely new. At its inception, writing was neither more nor less than the very quintessence of information—the classificatory aspect of language abstracted from the flow of experience and rendered visible. The origin of writing therefore constitutes, at one and the same time, the first information technology and the birth of information itself.

From its unique, informational origins, writing only gradually began to pattern itself upon speech. It did so at first in limited ways, making highly selective use of the words and syntax of speech in order to function more effectively in its capacity as information. In this form of early writing, the representation of speech was subordinated to the task of accounting. Only later, as writing began to break free of accounting and pattern itself more thoroughly upon speech, did it begin to constitute a new technology of communication. The early history of writing, then, tells the story of how an information technology evolved into the technology of communication we now take for granted.

Tokens, Emblems, and the Evolution of Accounting

The roots of writing extend back to counting, which is grounded in the ability to engage in "time-factored activity," an ability characteristic not simply of humans but of humanoids. The regular use of fire, dating back at least a half million years, entails the knowledge that dry wood burns better than wet and, hence, that fuel is most readily gathered during certain seasons. Likewise, systematic hunting is keyed to the knowledge of migrational patterns. These and other primordial activities necessarily involve an awareness not only of time's passage but also of temporal patterns, an awareness that long predates the appearance of Homo sapiens.

Judging from archaeological evidence, tallying, the precursor to counting, emerged after the arrival of Homo sapiens. Numerous bone and stone artifacts dating from the Ice Age, up to thirty thousand years ago, bear markings that apparently record the days between the phases of the moon. On some artifacts, these markings seem to correlate the phases of the moon with other natural patterns, such as the seasonal spawning of fish, the migration of animals, and the ripening of wild fruits and grains.

In its simplest form, this notational system consists of a series of notches clustered into groups, as on the "Ishango bone" from Central Africa, dating back around eleven thousand years (see plate 1). Analysis of this piece of petrified bone suggests that the groups of notches correspond to the phases of several consecutive lunar cycles. We can only speculate as to the reasons for keeping this record—whether for hunting, migrational, or ceremonial purposes—but it apparently represents an early instance of tallying, a form of record keeping that naturally led to counting.

The notches inscribed on the Ishango bone are "tokens," one of the forerunners of writing. In all likelihood, each notch depicts a finger, in the commonplace mode of finger counting, with fingers and notches as tokens for countable items. The system of counting by tokens rests on the principle that one token stands for one item in a group, without specifying which particular item or the nature of the group. For example, fingers can stand for days of the month, goats in a herd, or units of grain, with any finger standing for any countable item. This system is "token-iterative," for it forces the record keeper to count, "One item, two items, three items . . ."

Another forerunner of writing is the "emblem." Whereas tokens refer, generally, to any kind of countable item, emblems refer to specific objects. When we stop to think about it, emblems are all around us, in coats-of-arms, company logos, and national flags, to name only a few of the most obvious instances. The contemporary furor over flag-burning offers a faint reflection of the once powerful fusion of these symbols and the things they symbolized.

In their primordial beginnings emblems derived from totemic signs endowed with the qualities of the natural objects they represented. Thus the "Bear" clan had the bear as its totem and bear sign as its emblem, all three so infused with the quality of "bearness" — strength, courage, agility, speed — that they were essentially indistinguishable from each other. Note that the emblem of the "Bear" clan need no more resemble a bear than, say, the modern Chevrolet emblem resembles a car.

The earliest antecedents of writing, tokens and emblems merged to create more complicated systems of record keeping with the rise of settled agriculture and animal husbandry some ten thousand years ago. Under the inexorable pressure of accounting needs, emblems began to lose their totemic qualities, or at least have them substantially weakened, enabling symbol to be pried loose from thing symbolized. Symbols gradually became utilitarian devices and combined with token-iterative practices to create accounting systems in order to keep track of the diverse items prevalent in an agrarian economy. The evolution of accounting elicited the classificatory potential of language, culminating eventually in writing.

With their simple series of notches, the most primitive notational systems worked well for hunter-gatherers living from hand-to-mouth, but not for farmers storing produce. With the rise of settled agriculture in the area of western Asia known as the Fertile Crescent, the wider range of items characteristic of the new agrarian economy (sheep, goats, cattle, milk, grain, wool, bread, honey, fish) came to be depicted by pieces of clay with conventionalized shapes and markings (spheres, disks, cones, tetrahedrons, ovoids, rectangles, triangles; see plate 2). The shape of these clay pieces probably derived from the design of emblems that had assumed nonemblematic uses under the pressure of accounting needs. One could now indicate the size of a flock of sheep, not by notching a stick or counting pebbles, but with a given number of sheep symbols, as distinct from goat and grain symbols. This clarified the task of record keeping. Like the one it supplanted, the new form of record keeping remained token-iterative, in that the record keeper counted, "One item, two items, three items . . ."

A major step toward writing occurred soon thereafter, when the use of emblems became combined with that of tokens. This happened quite naturally as record keepers began notching or otherwise marking a single symbol to indicate a given number of items. A sheep symbol inscribed with five notches might stand for five sheep or some other figure, depending upon the means of counting sheep. Note, however, that regardless of whether a notch stands for one sheep or ten or sixty, it does not represent an abstract conception of number; rather, it simply embodies an object or collection of objects. A notched sheep symbol is still token-iterative, in that the record keeper essentially counts, "One sheep (or group of sheep), two, three . . ." The novelty of this development issues from combining tokens and emblems, previously used only in separate contexts, to signify the class and number of objects.

In this way increasingly sophisticated accounting procedures began to exploit what we have been calling the "classificatory potential" inherent in speech, which uses words to distinguish between classes of things, between names and their modifiers. The simple act of inscribing a symbol with notches points toward the eventual differentiation of pictorial and scriptorial signs, with the former conveying information holistically and the latter by classificatory means. As noted above, though, notched symbols remain token-iterative, and all such systems fundamentally pictorial, because they still depict each item or group of items recorded.

At this point the skeptic may again raise objections, for intuition tells us that tokens and emblems already constitute information, long before the invention of writing. Bones notched to record the phases of the moon, different collections of clay symbols, and symbols covered with notches—all this, common sense says, amounts to data. That new systems of record keeping evolve under the pressure of accounting needs only reinforces this intuition.

But our intuition is mistaken. We cannot overestimate how counterintuitive the preinformational world is for us literates. We misconstrue it because of our ingrained reflex to separate the knower from the known, to thrust mental objects mediately between ourselves and the world. Information's invention, writing, causes this reflex. Our information-bound intuition leads us into the error of anachronism, the automatic tendency, revealed in the previous chapter, to view orality in the mirror of literacy.

In order to understand the birth of information, we need to switch perspectives and imagine record keeping from an oral frame of mind, where speech is continuous with the flow of experience. The functional counterpart to this flow, token-iterative accounting remains embedded in the practical activity of keeping track of particular things. The physical presence of sixty sheep symbols—or of a single symbol with sixty notches—simply depicts the concrete phenomenon of sixty sheep. It serves as a visual map, a mnemonic device, organizing experience unreflectively (that is, without separating knower from known). The token-iterating record keeper has not detached himself from this concrete experience sufficiently to elicit information from it.

Information derives from the twofold process of abstraction involving the related movements of "drawing away from" and "taking out of." When he draws away from the flow of experience in a consistent manner, the record keeper will come to realize that he expresses the phenomenon of sixty sheep verbally in two words—a noun and an adjective, a name and a number. This point marks the threshold of writing, and once crossed, the mind begins to reflect on its own products, to mediate experience with mental objects.

From Token Iteration to Emblem Slotting

For over five thousand years, clay symbols were commonly used in token-iterative accounting systems across western Asia, from modern Turkey to Pakistan. Around the fourth millennium B.C., they began to change as the rural subsistence economy gave way to an urban one based on trade. Urbanization developed rapidly in the region of Lower Mesopotamia (in modern Iraq) because of the exceptional fertility of the alluvial plain between the Tigris and Euphrates rivers. The southern portion of this region, occupied by the Sumerians, became the most densely settled.

Here flourishing trade led to the transition from token-iterative to what anthropologists call "emblem-slotting" systems of record keeping. These systems exploit the fact that language classifies, that the words "sixty sheep" classify or "slot" by name and number. When translated into a new system of record keeping, this realization facilitates keeping track of the wider range and growing number of goods in a burgeoning economy. With the emergence of emblem slotting, we cross over from picture drawing to writing, from experience to information.

As an urban economy arose in Mesopotamia, the existing token-iterative accounting system accommodated the wider range of goods by simply expanding the different types and subtypes of clay symbols. It had greater difficulty, however, representing the larger numbers of items traded. We can easily imagine keeping track of twenty sheep with twenty sheep symbols, but what about two hundred sheep, or two thousand? Economic growth led to "bundling," which evolved from the practice of notching symbols, whereby one kind of symbol came to stand for a fixed number of goods of that type. Shepherds could thus keep track of the size of a flock by means of different kinds of sheep symbols, with one kind denoting single sheep, another ten sheep, and another sixty. They could denote larger flocks by taking the symbol for sixty sheep and inscribing it with a drawing of the "ten-sheep" symbol, thereby indicating six hundred sheep.

Absence of clear evidence keeps the details and chronology of these advances in accounting very murky. Apparently, different accounting systems existed in different Sumerian cities, and within each city different systems were used for counting different items. Sometime between 3500 and 3100 B.C., local systems began to coalesce into regional ones. In Lower Mesopotamia, for example, the different systems for counting sheep, goats, and cattle became consolidated into a single system for livestock. (This system still differed from that used to count dead animals, or fish, or dairy products, or measures of grain, or units of the calendar.) Consolidation in accounting systems likely encouraged the next, short step: from inscribing a sixty-sheep symbol with a ten-sheep sign to inscribing a clay tablet with the sign for a sheep, followed by signs from the livestock system indicating the number 600.

This point marks the critical transition from token iteration to emblem slotting. The new accounting system differs from its predecessor in two ways. First, it clearly distinguishes between the name and quantity of items, each having a separate "slot" in the entry. And second, it indicates quantity not by a series of iterative notches or symbols but by a sign for a number. The number system is generalized enough that the sign itself does not specify exactly what is being counted—for example, the livestock number system does not distinguish between sheep and goats. The record keeper thus must name the items in question and then modify them, as in the example "sheep, 600."

At this juncture, the record keeper has realized graphically the classifying potential of language, in an invention distinct from both picture drawing and speech. On the one hand, he has broken the isomorphism between pictures and objects. We cannot count the signs of "sheep, 600" in a token-iterating fashion because they bear no pictorial relation to six hundred sheep. On the other hand, his item and number signs bear no phonetic relationship to the spoken language. They are purely semiotic.

Writing derives from the fact that language classifies. Under the pressure of accounting needs, this classificatory feature was abstracted from speech, becoming the basis for an entirely new form of record keeping that, by definition, served less a communicative function than a mnemonic one. As a visual aid in remembering, writing differed fundamentally from other visual aids, such as maps, which despite their abstract qualities still served to mirror reality. Unlike maps and holistic forms of record keeping based on picture drawing, writing did not so much mirror the objects of the world as translate them into a different set of objects—quantitatively and qualitatively—bearing no pictorial resemblance to the world. These objects, the contents of mind and memory separable from experience, comprise information.

How could one become aware of the classificatory aspect of language in an oral culture? We can only speculate, but it seems reasonable to assume that this awareness derived from reflection upon the instruments of commemoration, the narratives by which the culture constituted itself. Swirling around us like air, everyday speech is an unlikely object of reflection because it obliterates its own traces in an ongoing flow. Highly memorable, rulelike statements—"Don't stand up in the boat!"—constitute a possible exception. But in the oral culture this statement would have only fleeting existence apart from the practice of boating, where the message probably overpowered the medium.

Such is not the case with speech designed to sustain social consensus, where the medium of commemoration receives special attention. Eric Havelock has aptly described this medium as an "enclave of contrived speech," reserved for special purposes. It is intentionally archaic, as removed from everyday usage as Shakespearean English is from our modern idiom. We have disputed Havelock's contention that this contrivance serves the purpose of information storage, showing instead that it preserves a series of commemorative, visual images both sustained and sewn together by acoustic formulas. An automatic communal process of abstraction produces these generalized images and formulas, having reduced them to a common denominator without separating knower from known. Yet, even though the members of the audience engage uncritically in the act of commemoration, they remain nonetheless highly attuned to the story and the medium of its conveyance, such that they can recognize (and, if necessary, reject) departures from established models.

The enclave of contrived speech thus already stands in relief as the focus of considerable attention, increasing the likelihood that someone in the community might begin to reflect upon it further, consciously, "drawing away from" it enough to see how it functioned. Recall that in the case of the Greek epics, the enclave was characterized by verbal formulas, most notably the "noun-epithets," which combine a given set of names with a given set of modifiers. As one "draws away from" these formulas, one perceives not merely specific information about the names and modifiers—that "Achilles" is "swift-footed," "strong," "brave," or whatnot—but the general notion that speech functions, among other ways, by "adding to names."

Of course, we cannot extrapolate directly from the Greek oral experience to the Mesopotamian one, where the enclave of contrived speech was constituted differently. But the basic point is clear: The activity of commemoration in Mesopotamian oral culture probably encouraged conscious reflection upon the nature of language, revealing its classificatory potential. The genius of the invention of writing lies in abstracting this capacity from speech, in using it to create something different from speech, something totally new: information.

The Evolution of Writing from Information to Technology of Communication

The earliest writing is called "proto-cuneiform" because it precedes cuneiform writing, whose name comes to us from the Latin *cuneus,* or "wedge," and describes the shape of symbols impressed in soft clay by a stylus with a triangular-sectioned tip. Instead of impressing it with a triangular-tipped stylus, scribes incised proto-cuneiform with a pointed one (see plate 3). Archaeologists have unearthed the oldest examples of this type of writing, dating roughly between 3200 and 3100 B.C., at the site of the great temple in the Sumerian city of Uruk. The temple served both a religious and an administrative function, as either the owner or the redistributor of agricultural produce. Writing may very well have originated here, at this site, as the expanding urban economy strained prehistoric accounting.

Proto-cuneiform tablets are pure information, abstractions from the flow of experience without any concession to the exigencies of communication. The officials involved already knew the nature of each transaction, so they only needed to record the type and amount of goods changing hands, and sometimes the date. These entries mix "pictographs" (objects drawn in whole or in part), symbols (such as a bundle of reeds representing a goddess), emblematic signs (such as a circle inscribed with a cross, denoting small livestock), and signs from the appropriate number systems. Such markings bear no syntactical relation to each other, nor are they derived phonetically. Rather, they simply encode information, the context of which the officials took for granted. (The code says nothing about the context, thus we can only surmise the actual nature of these transactions.) And when this information became outdated, with each harvesting of fields or flocks, it was literally "dumped," old tablets becoming landfill.

Once they devised the basic signs of proto-cuneiform writing, scribes could modify and combine them, creating new signs to inform larger portions of experience. When they inscribed the sign for head with parallel lines, they formed the sign 'mouth'. Also, they could combine signs to depict ideas for which no single pictograph, symbol, or abstract sign existed. The sign for head, when combined with the sign for ration (a bowl), yields the sign 'disbursement' (see plate 4). By such means, the repertory of signs expanded considerably, but it nonetheless remained limited to *things*, without expressing relationships between them. It remained, in other words, pure information.

Writing began to model itself upon speech not in order to represent language but rather in order to inform still wider reaches of experience. About one hundred years after the origin of writing, phonetization makes its appearance, adding even greater flexibility in the creation of new signs, especially those denoting different individuals and diverse goods. The need to identify individuals unambiguously may have spurred the development of phonetic signs. Originally, officials and others marked their transactions with personal seals. As the economy expanded, along with the size of the bureaucracy overseeing it, the number and complexity of these seals mushroomed. Phonetization may have originated as a method of simplifying the task of personal identification. Perhaps the earliest example of phonetization is found in the formation of a personal name, rendered in English as "Enlil-gives-life." It consists of symbols denoting the god Enlil combined with a sign depicting an arrow. (The Sumerian words for arrow and life are pronounced similarly.)

In this appellation, the word for arrow matches the sound for part of the name, transforming the object sign into a phoneme. Such a sign is called a "rebus." To illustrate rebus writing with an example from English, imagine combining the picture of a bee with that of a leaf and expressing the result phonetically as two syllables, *be-lief,* itself an incorporeal concept. The Sumerian language offered fertile ground for the development of this device. It is an "agglutinative" language comprised of base words, with prefixes and suffixes denoting syntactic relations. The base words are largely single syllables with a presumably high percentage of homophones, offering a number of applications for any given rebus. The very nature of the Sumerian language thus facilitated the use of the device, which gradually evolved into a syllabic sign, depicting a sound rather than a thing.

Although their language fostered use of the rebus, even the Sumerians had relatively few of these devices, which they necessarily had to use in various combinations. We might wonder how they could write a monosyllabic base word with signs depicting several sounds, but this merely reflects our alphabetic and literate bias, in which phonetic syllables are clearly differentiated by spellings enshrined in "the dictionary." At the inception of phonetization, however, the notion of a syllable was quite flexible, if not completely arbitrary. The Sumerians had no difficulty writing a monosyllabic word, such as *bal* ("to dig") with two separate syllabic signs, *ba* and *al*.

Concurrently with phonetization (c. 3000 B.C.), proto-cuneiform script gave way to cuneiform proper, as scribes began using a stylus with a triangular-sectioned tip in place of the original one with a pointed tip. Wet clay was not well suited for drawing. Lines could become easily clogged with excess clay and, as the tablet dried, the sharp edges of the inscription could crumble away, making it unreadable. Consequently, scribes turned to the far more suitable, triangular-tipped stylus, an ideal shape for the practice of impressing wet clay because it makes clear, directional punch marks, thick at one end and thin at the other. The triangular-tipped stylus rendered the curved, proto-cuneiform pictographs more abstractly, and eventually they lost all remaining resemblance to the drawings from which they had originated (see plate 5).

Scribes wrote early proto-cuneiform from top to bottom, in columns running from right to left. In later versions of this script, they wrote the pictographs at a 90 degree angle from the vertical, in rows running horizontally from left to right. (When they read the tablets, they rotated them back to the vertical.) The reasons for this change are obscure, perhaps reflecting a shift from small, square writing tablets to larger, rectangular ones, which had to be held with the fingers rather than in the palm of the hand. Eventually, after cuneiform had fully supplanted proto-cuneiform, scribes read the signs in the same direction as written, for the growing abstraction of the script made rotating the tablet back to the vertical unnecessary. (For the evolution of cuneiform script, see plate 6.) These advances in abstraction complement the spread of phonetization, which steadily transformed object signs into syllabic signs.

As phonetization progressed, writing further modeled itself upon the syntax of speech, which enhanced its capacity for information storage. We have difficulty deciphering early proto-cuneiform tablets because they do not clarify the relationship between the entries for each transaction. Later tablets, however, make selective use of words and syntax to record a greater amount of information. For example, one entry translates roughly as "40 kagu-breads baked at the rate of 50 per bán" (where "kagu" denotes a type of bread and a "bán" equals about ten liters of grain). This kind of entry stores a great deal more information than earlier records, which could only denote loaves of bread or units of grain in separate entries, with no reference to whether they were baked or harvested, let alone the rate of the activity. Even so, these later tablets still do not record speech, for the terms they use and the syntax they follow are highly specialized for accounting purposes. Writing, to reiterate, evolved from pure information into a technique of information storage before it became a full-blown technology of communication.

That technology emerged only when Mesopotamian writing began to break free of accounting constraints and model itself more thoroughly upon speech. This development occurred toward the middle of the third millennium, when the Sumerians applied their script to the language of the Akkadians, a Semitic people who had settled the northern half of Lower Mesopotamia. Unlike the Sumerians, who seem to have migrated to the region in one wave, the Akkadians had steadily infiltrated over the centuries, gradually making their influence felt not only in the north but also in the south, where they increasingly mixed with the Sumerians. In Sumerian cities with a growing Akkadian population, scribes confronted the need to render Akkadian proper names in Sumerian script for accounting purposes. And as trade between the northern and southern parts of the region flourished, Semitic names for goods also found their way into Sumerian records. Eventually, the Akkadians themselves began using Sumerian signs to transcribe their own spoken language.

Whereas Sumerian is an agglutinative language composed of monosyllabic base words, Semitic languages like Akkadian are "inflected." They denote syntactic relationships, not by adding prefixes and suffixes to base words, but by declining nouns and conjugating verbs, by the phonetic modification of words that are, consequently, polysyllabic. The Sumerians used their monosyllabic signs to transcribe polysyllabic Akkadian names and words phonetically. And the Akkadians themselves expanded this practice when they eventually learned the art of writing, using Sumerian script as a means of expressing their own language. A fortuitous blending of the two language groups in the region thus boosted the process of phonetization, enabling writing to break free of information storage and emerge as a new technology of communication.

This process culminated with the rise of the Akkadian empire around 2300 B.C., when the Akkadians established their political hegemony over the Sumerians. By then, writing had become sufficiently modeled upon speech to lead to the development of a distinctive literary tradition, which assumed the peculiar form not of continuous texts, what we call "literature," but rather of lists. We shall now turn to these lists, for they reveal how the newly emerging technology of communication redoubled the classificatory potential of language. The pictographic form of Mesopotamian script encouraged a distinctive kind of list making, specifically designed to bring order to the avalanche of information created by writing.

Literacy and List Making in Mesopotamia

Although a late Assyrian collection dating from around 650 B.C., the royal library at Nineveh gives us a glimpse of a Mesopotamian literary tradition preserved religiously by almost two millennia of scribes. Of the seven hundred extant cuneiform tablets in the royal library, the largest group consists of three hundred "omen texts." These lists consist of short, conditional statements correlating various phenomena with predicted outcomes, such as "If a man's chest-hair curls upwards: he will become a slave." The next largest group, two hundred tablets, comprises sign and word lists organized according to a wide variety of principles, largely to aid Akkadian-speaking scribes in mastering Sumerian, a dead language early in the second millennium. A related group of one hundred tablets provide interlinear translations of Sumerian incantations and prayers. Finally, some one hundred tablets contain assorted conjurations, proverbs, and fables, of which only about forty record epics, like the famous story of Gilgamesh. Even taking into account the loss or destruction of tablets, continuous texts—what we call "literature"—constitute only a tiny percentage of the Mesopotamian scribal tradition, which consists chiefly of lists.

To some extent the nonnarrative nature of this literary tradition reflects the requirements of training in a difficult and obscure scribal art, with its emphasis on word lists and copybook phrases, on learning a dead language by rote. But the preponderance of lists also indicates that, as the embodiment of the linguistic propensity to classify, writing tends to encourage further classification. As long as writing remained limited to record keeping, as either pure information or, at most, a technique of information storage with restricted terminology and syntax, the tendency to classify was tightly constrained. But when writing broke free of record keeping and became a technology of communication modeled upon speech, the classifying urge exploded.

To illustrate how and why this happened, recall our earlier example of the classifying nature of language. When we say "red block," we classify an object by its shape and color. Writing more fully exploits this aspect of language, literally abstracting these classes, fixing them apart from the flow of speech and transforming them from evanescent sounds into visual objects. Their almost palpable presence heightens our awareness of them as classes, prompting us to list different color blocks (blue, green, yellow) and different red objects (ball, flower, apple), tempting us, in other words, to fill out the newly perceived categories of experience. By drawing attention to the classifying possibilities of language, writing encourages the phenomenon of list making.

The word 'list' is so commonplace we hardly give it a thought. From the Old English *liste,* "hem" or "border," the term formerly meant a strip of cloth, specifically "the selvage, border, or edge of a cloth, usually of different material from the body of the cloth." It also denoted border in the sense of "boundary," specifically in its plural form as "the palisades or other barriers enclosing a space set aside for jousting," whence the phrase "to enter the lists." The implication of boundaries carries over into our contemporary usage of the word as "a series of names or other items written or printed together in a meaningful grouping or sequence so as to constitute a record." In contrast to the flow of speech, lists create boundaries, which both distinguish the individual entries within the list and separate all its items from those outside. These internal and external boundaries encourage the scrutiny of its entries, individually and as a group, revealing the possibility of new classifications.

The Mesopotamian lists devoted to scribal training give free rein to the impulse to classify. In addition to Sumerian sign lists, and bilingual lists translating Sumerian terms into Akkadian, we find a profusion of topically arranged word lists. According to the noted Assyriologist A. Leo Oppenheim, one such list encompasses topics like "trees, wooden objects, reeds and reed objects, earthenware, leather objects, metals and metal objects, domestic animals, wild animals, parts of the human and animal body, stone and stone objects, plants, fish and birds, wool and garments, localities of all description, and beer, honey, barley, and other foodstuffs." Another list presents various classifications for human beings, under such headings as officials, craftsmen, and cripples. Beginning as a pedagogical device for teaching Akkadian scribes how to write and pronounce Sumerian words, the activity of list making transformed itself into a topical compendium of all received knowledge.

Of course, oral cultures make lists too, as illustrated by tribal genealogies, as well as more complex lists like the "Catalog of Ships" in book 2 of the *Iliad.* This list, which most scholars regard as part of the oral record handed down to Homer, describes the peoples, geography, and rulers of Greece, as well as the numbers of ships and men in the expedition to Troy. Unlike Mesopotamian word lists, the Homeric one is storied, narrating actions: Over the various peoples of Greece, kings are ruling; in the various regions of Greece, sheep are pasturing; and onto the ships from each region, men are embarking. Commemorative patterns of sound and image underlie these concrete situations, subordinating oral list making to community ritual and embedding it in a kind of liturgy. By contrast, writing frees the mind from the mnemonic necessities of oral culture, allowing its energies to flow naturally in the direction of heightened classification. Writing does not so much *create* the urge to classify as *liberate* it, encouraging new forms of intellectual play.

Common to all members of the human species, the activity of play assumes a formative role in each of our information ages, as heightened abstraction calls forth the desire to elaborate and master the resultant new categories of knowledge. The great Dutch historian of culture Johan Huizinga defines play as "a voluntary activity or occupation executed within certain fixed limits of time and space, according to rules freely accepted but absolutely binding, having its aim in itself and accompanied by a feeling of tension, joy and the consciousness that it is 'different' from 'ordinary life.'" Play, for Huizinga, is fundamentally agonistic, in the Greek sense of a contest where participants strive with each other to demonstrate their power or skill.

Although the Mesopotamians did not have an agonistic, warrior culture like the Greeks, we can still see the play element at work in their list making, as they strive to demonstrate their mastery over the information they have created. Indeed, list making offers exceptionally fertile ground for the growth of play. Lists constitute a well-defined, rule-bound arena in which the activity of naming takes place for its own sake, accompanied by a feeling of tension manifest in the urge to be exhaustive. Of the Sumerian word lists, one modern scholar observes: "Every group of names, be it of stones, fields, or officials, is complete, perhaps too complete. In some instances the list comes to resemble some of the modern student exercises the chief aim of which is to compose as many sentences as possible with any given noun." The urge to be exhaustive derives from a form of intellectual play that seeks to organize and control the newly perceived categories of experience.

The very script that made the activity of listing possible reinforced this urge and its attendant play element. Although writing in Mesopotamia became increasingly phoneticized, it never disassociated itself completely from its pictographic origins. Instead, phonetization simply added another layer of meaning to script, existing alongside the already established one. An object sign came to denote both a syllable *and* an object. And these readings could coexist, giving the script a peculiar characteristic, epitomized as a kind of linguistic "polyphony." The dual nature of the script accentuated the play element in the Mesopotamian practice of "word divination."

Recall that the early stages of phonetization harbored considerable flexibility in the syllabic spelling of words, but over time spellings tended to become fixed by convention. Copyists, solely concerned with the pragmatic use of writing, devised conventional spellings for a whole range of entities—gods, kings, officials, corporations, animals, and inanimate objects. But another group of scribes, scholars concerned with the "meanings" of words, utilized not only conventional spellings but also all conceivable syllabic ones to unpack the cosmic import of words assigned by the gods.

In the divination of a word, each possible syllable had at its core the Sumerian sign for a monosyllabic word (to say nothing of the possible homophones for that monosyllable), which derived from a pictograph representing a real entity in the world. For example, the god Marduk was known by the name "Asari," among many others. In spelling the name phonetically, copyists customarily rendered it by means of the syllables, *a - sa - ri;* but scholar/scribes did not feel so constrained, dividing it into, among others, the syllables *a - sar - ri.* The syllabic sign for *sar* depicts two ears of grain along a furrow, from which the scholar/scribes divined agriculture as one of Marduk's attributes. Additional attributes inhered in the remaining syllables, their homophones, and other syllabic spellings of the name.

The scholar/scribes thus revealed each word to consist of many words whose pictographic values encompassed the qualities inherent in the thing named. With singular intensity and determination, they analyzed words into every conceivable syllable. And when they exhausted the meaning of syllabic signs, they moved on to examine the homophones for each sign. The "polyphonous" nature of words made the analytical possibilities virtually endless, spurring the scholar/scribes to ever more ingenious readings.

The very nature of its script intensified the play element in Mesopotamian scribal culture, an element that manifests itself in exhaustive list making. But play alone cannot account for this phenomenon. The practice of word divination indicates writing's sacral quality. Although religious sentiment can express itself through forms of play, encounters with the sacred are not playful at heart. When considered seriously, the "Almighty" obliterates all playing fields, along with the puny feelings of tension and joy associated with them. Mesopotamian writing is shot through to its core with this sacral quality, which ultimately circumscribes the play element and transforms the fullest possible enumeration of entities into a duty.

All writing can be imbued with the sacred—we need only recall the significance of the Greek letters alpha (A) and omega (Ω). But whereas Christians regard these letters as symbols standing for something outside the Greek alphabet, the Mesopotamians found the wellspring of the sacred within the very signs themselves. Word divination reveals the "logocentrism" of the Mesopotamians, as distinguished from our "reocentrism." In our reocentric perspective—from the Latin *res,* or "thing"—reality is embedded in things, from which words derive their meanings. In the logocentric view—from the Greek *lógos,* or "word"—this relationship is reversed, with reality being embedded in words, to which things must conform. From a logocentric perspective like that of the Mesopotamians, someone named "Trudy" would never tell a lie (even though her full name, "Gertrude," has nothing to do with honesty) and a wimp named "Musselman" had better like shellfish.

Sumerian writing embodied the fundamental nature of things in the pictures that made up the words. Indeed, by means of naming the gods assigned reality to things that otherwise would not exist; reality was something divinely inscribed. No wonder, then, the scholar/scribes strove to be as exhaustive as possible in the analysis of words. Although they may have contested with each other for increasingly ingenious readings of words, their all-inclusive lists ultimately have the compulsive quality of sacred ritual.

Nowhere is this quality more apparent than in the "omen texts," topically arranged, short entries taking the form of "if . . . then" statements, as in the example, "If a man's chest-hair curls upwards: he will become a slave." Though the modern mind strains to detect any causal relation here, the omen may have originally derived from observation, when a king once subjugated such a race of men. Other omens derived from a kind of word divination, in which the will of the gods lay inscribed, either in the words of an omen or in the phenomenon described, whose physical configuration was likened to a divine pictograph in the secret script of the gods.

Exhaustive in scope, the omen texts encompass (according to the French Assyriologist Jean Bottéro) "stars and meteorites; the weather and the calendar; the configuration of the earth, of waterways, and of inhabited areas; the outlook of inanimate and vegetal elements; the birth and the conformation of animals and their behavior, especially of man himself—his physical aspects, his behavior, his conscious and sleeping life, and so on." Not only are the topics exhaustive but so too are the enumerations within each topic. Regarding dreams, for example, the omen texts cover a wide range of subjects: dreams about voyages, about producing various items, about sending and receiving goods, about consuming food and drink, about various bodily functions. And each subject, in turn, is broken down into every conceivable variant. Dreams about micturition encompass, among others, instances of urinating on a wall, on a wall and a street, on several streets, on reeds, onto an irrigated field, into a river, into a well, sitting down, up into the air. Obviously, one instance leads to another, as the scribes attempted to elaborate all the dream possibilities systematically.

From our point of view, the omens they portend—"If he urinates upwards, he will forget what he has said"—derive from various forms of symbolic interpretation. If wetting oneself is a sign of senility, such a dream may well augur forgetfulness. The Mesopotamians, however, regarded dreams as divine pictographs, written either in the secret script of the vision itself or more conventionally in the pictographic words describing the vision. In the latter case, words might convey the meaning of an omen through their constituent pictures or through their phonetic similarity to other words, which themselves might be susceptible to further analysis.

Because reality was fundamentally something inscribed, one could make a phenomenon real simply by expressing it in words, ignoring what we reocentrists (burdened, by contrast, with the things of the world, with what we call "reality") would regard as its plausibility. Concerning sacrifices, for example, we find an omen covering the possibility of a liver with two gallbladders, an event rare enough to be "ominous" indeed. Once having established the category, the scribes felt obliged to expand it, listing individual omens for livers with up to seven gallbladders. Comparably, if a woman gives birth to twins, one omen is specified; if triplets, another; if quadruplets, another; and so on, up to nine babies! Thus the activity of list making took on tremendous momentum, filling all the actual and imagined categories of experience, which attained equal reality in the script that depicted them.

We observed above that writing does not create the urge to classify but liberates it, actualizing the potential inherent in language. And we saw how this activity naturally partakes of a play element common to the human species. By now, we can clearly perceive that each culture realizes this potential in its own distinctive way.

The sacral quality of Mesopotamian writing, rooted in its pictographic origins, channels the tendency to classify in a particular and (to our sense) strange direction, epitomized by exhaustive list making. Each list consists of a multitude of systematically elaborated categories. But all this systematizing serves not to classify the things of the world in the kind of hierarchical pyramid we reocentrists would expect, one that analyzes reality by genus and species, reflecting the order of nature. Instead of classifying in this way, the categories exist chiefly to focus the mind and thereby to aid it in a thorough naming of all possible items and phenomena, for if one can name something, one can know it, not as part of the natural hierarchy but purely in itself. The systematic quality of Mesopotamian list making, then, evinces not so much a vertical as a horizontal organization of knowledge, an ever-expansive naming of all the possible phenomena within each category. The Mesopotamians structured knowledge in an additive rather than analytical manner.

The Anatomy of a List

Up to now we have freely used the term 'classification' and its cognates to describe the conceptual matrix within which writing originated and from which it evolved. But, as intimated above, we should guard against being too Hellenic, for the term is freighted with the analytical or, better put, taxonomic assumptions of the Greek enlightenment, a period almost as far in the future from the Mesopotamian era as it is in the past from our own. Whereas the Greeks would climb a ladder of abstraction, ascending step by step from the flux of the sensory world to the empyrean, first principles of existence, the Mesopotamians stopped after only one or two rungs. Let us now dissect an actual list to see how the pictographic origins of Mesopotamian script, and the accompanying urge to be exhaustive, defined and circumscribed the natural impulse to classify, and thereby hindered the kind of hierarchical thinking at which the Greeks, scions of alphabetic literacy, would later excel.

A list of lists, the so-called "Sumerian King List" presents an ideal specimen for our purposes, revealing the full extent of the classifying impulse excited by Mesopotamian literacy. It survives in several variants which, taken together, provide an ongoing record of Mesopotamian rulers from the legendary time before the Flood to about 1800 B.C. Philological analysis has revealed that all the variants stem from a common source, composed around 2100 B.C. The King List originated during a period of national revival under the leadership of the Third Dynasty of Ur, when the descendants of the Sumerians regained control of Mesopotamia. Pride awakened by this development gave rise to the notion that all Mesopotamia had always been under a single rule, now finally restored to the proper hands. The King List both sanctions and memorializes this idea.

In contrast to the tales of Homer, the list is clearly of literate origin, being so spare as to provide no story for the memory to grasp. The bulk of the list, running over four hundred lines, reads according to the following simple formula: "B, son of A, reigned X years; C, son of B, reigned Y years." The list of reigns is punctuated every few lines by a conventional transition, generally one stating that "city X was smitten with weapons, and its kingship to city Y was carried," followed by the next enumeration of reigns. In some instances, the list contains anecdotal information concerning the origin and achievements of a ruler, as in the following:

> Etana, a shepherd, the one who to heaven
> ascended,
> the one who consolidated all lands,
> became king and reigned 1,560 years.

This is one of the most elaborate of these narrative flourishes (if we can call them that), most being limited to a brief mention of the native city or profession of a future king. In short, there is no "story" here, at least none that the oral imagination can lay hold of, with its reliance on heroic action and concrete detail.

Not only is the list a purely literate creation, but it derives from written sources. Philological analysis discloses two kinds of antecedents, corresponding to the two kinds of materials contained in the list. The names and reigns of kings trace to the "date lists" of various cities, enumerations of local rulers and dignitaries extending back to the very remote past. The anecdotal information about kings descends from a host of oral epics and legends preserved in writing. For instance, the above quoted information about Etana derives from the "Etana" epic, probably written down long before the author of the original King List set to work. He thus faced the task of organizing in a single compendium a wealth of information drawn from two very different kinds of documents.

That the date lists would obviously be his primary source, with epic literature restricted to a supplementary role, somewhat eased his conceptual burden. But before using the date lists, he had to find a means of unifying their diverse accounts. Each of these lists commemorates the independent political tradition of a city, generally ruled by a king, that existed in competition with other cities in the region. The ruler was regarded as having been chosen by the god of the city, and struggles between cities were viewed as conflicts within the assembly of gods. Our scribe, however, wanted to commemorate the Sumerian revival, when all Mesopotamia came under the control of a single dynasty in Ur. This phenomenon of a single city's dominance over the entire land required that he devise a new interpretation not only of political power but also of its theological basis, enabling him to merge his sources into a single account.

The scribe's solution to this problem demonstrates his considerable ingenuity, if not genius. He cleverly modified the traditional formula used in the date lists to describe the divine origins of a city and its rulership. The boldness of his departure from tradition, apparent in the opening lines of the King List, is masked by the flat, repetitive, dogmatic style of Mesopotamian writing: "When the kingship was lowered from heaven / the kingship was in Eridu." In place of separate kings, each ordained by the particular god of a city, he has posited a single, divinely ordained "kingship," transferred from one city to the next by an assembly of gods acting in unanimity.

Having found a means of combining the date lists, the scribe could arrange them sequentially, hence charting the course of a kingship that would appear to pass from one city to the next. This posed an added difficulty, though. His other body of sources, epics and legends, revealed that some of the rulers of different cities were contemporaries. The kingship could not reside concurrently in different cities. More so than the preceding problem, this one no doubt taxed our scribe's critical abilities to the utmost.

Logically, he might have resolved the conundrum by determining at what point the kingship was transferred from one city to the next, utilizing only the portion of each date list corresponding to the presence of the kingship in that locality. Although this solution may seem obvious to us, the scribe found it unacceptable because it violated the spirit of Mesopotamian list making, which tends toward completeness, toward the inclusion of all material appropriate to any given heading. Once named, all the rulers of the date lists had attained sufficiently reality as rulers to require their inclusion in the master list, lest it be incomplete. The need to be exhaustive further exacerbated chronological problems, for the scribe could not arrange the unexpurgated date lists successively without creating lengthy chronological separations between contemporaneous rulers.

Given the constraints imposed upon his thinking by the need to be exhaustive, our scribe devised what must have seemed to him a brilliant solution, however imperfect in our eyes. He arranged the material of the date lists sequentially, yet so as to minimize chronological disparities. Dividing each date list into its constituent dynasties, he interspersed the latter to form the King List in such a way that contemporaneous rulers became near contemporaries. He thus preserved completeness without egregiously violating chronology.

The Sumerian King List demonstrates how writing fostered critical abilities by accentuating the classificatory potential inherent in language. Although the members of an oral culture could conceive the class 'kings'—drawn from their experience, direct or vicarious, of individual rulers—it would exist only as a passing thought. For sustained thinking about kings, they relied on the activity of commemoration, the telling of stories about the heroic actions of specific figures.

Recall that an oral culture retains these stories in its collective memory by means of generalized images to which all members of the community can assent. Although these images are abstracted from the narrative action of the story, they are not themselves abstract; they preserve the generalized image, not of "a king," but of "the king." Generalized and stripped of local detail, the image is nonetheless specific to the story it serves to map.

Writing enables one to do readily what one cannot otherwise so readily do, create a class of "kings" standing apart from the stories told about them. Such is the case with the date lists for individual cities. Their initial, mythological sections, recounting the divine origins of each city, are no doubt abstracted from stories and legends, whether oral or written. The later sections, constituting a quasi-historical enumeration of rulers, are drawn partially from legend, partially from memory, and partially from records.

The author of the Sumerian King List could build upon this initial, classifying effort in two ways. First, he could compare the collected date lists with the collected epic literature, establishing a web of chronological relations between the rulers of different cities. Second, motivated by this new information, he could subdivide the date lists into their constituent dynasties. The latter innovation grows directly out of the activity of list making, which encourages the scrutiny of items, both individually and as a group, and reveals the possibility of new classifications. Writing thus stimulates an expanding range of critical activities whose complexity reaches beyond the confines of an oral culture.

Yet the nature of the script constrained the very critical abilities it spawned. Modern commentators generally regard the King List as embodying an implicit or nascent concept of kingship. The term 'concept' comes from Greek thought, the product of advances in classification associated with the coming of alphabetic literacy. It derives from a process of analysis that breaks the whole into parts and establishes a hierarchical relationship between them. The hierarchy is predicated on the ability to "draw away from" one's abstractions, to view them from a perspective that generates new abstractions. For our purposes, then, conceptualization is the process of abstraction operating on its own products, creating different kinds of information at each new level of abstraction.

At first glance, the kingship seems to be a concept. From the mass of kings in the collected date lists, the scribe appears to have abstracted the dynasties, and from the dynasties, the kingship. Were this a sequence of hierarchical abstractions, ascending from the many instances of kings to the singular idea of kingship, it would have yielded different kinds of information at each level. Yet we find that the information at the level of the collected date lists is exactly the same as at the level of the dynasties, which is the same as at the level of the kingship—each contains all known kings. Far from being a concept or class in the Hellenic sense, the kingship is simply an additive category invented to facilitate the more inclusive naming of kings.

The classifying impulse in Mesopotamian writing was thwarted by the urge to be exhaustive, an urge that derived from the pictographic origins of Sumerian script. Despite the growing role of phonetization, the Mesopotamians could never transcend the belief that the nature of things was divinely inscribed in their words. This belief led them to channel their classifying activity not toward higher and higher abstractions but toward an ever more thorough naming that ultimately sought to fix the nature, and thereby the knowledge, of the cosmos.

The pictographic origins of Sumerian script define the essence of information in ancient Mesopotamia. It originated as writing, the visual expression of language's ability to classify. As further manifestations of this ability were increasingly modeled upon language, writing unleashed the urge to classify the information it had created. And once freed, this urge acquired a momentum of its own, operating upon its own products, which naturally became objects of further scrutiny. The Sumerian King List—a list of lists—exemplifies this tendency. But Mesopotamian list making, though generated by a classifying urge, did not ultimately entail classification.

Of course, the lists were organized by categories, but these existed primarily to facilitate the exhaustive naming of entities and only incidentally to classify them. In fact, the real relations between entities were not classificatory at all; rather, they were inherent in the written names of things, whose constituent signs and their cognates depicted other things, without any obvious connections apparent to the modern mind with its Hellenic predisposition. The very nature of Sumerian script thus encouraged a sequential decoding of names, revealing the myriad pictographic tendrils through which vital sap flowed to animate the things of the world.

We shall now turn to a consideration of the Greek alphabet, which broke the pictorial tie between words and things, removing the barrier to classification. In so doing, it permitted the process of abstraction to operate freely upon its own products, giving rise to different kinds of information at each new level of abstraction. Information thus conceived was not synonymous with the world but rather constituted a body of facts about it. The activity of play accentuated this process of abstraction, yielding a hierarchical vision of the order of things, an order inherent in the things themselves rather than in the words used to express them.

PLATE 1. *Three views of the "Ishango bone."* Analysis of the bone's three columns of notches suggests that the columns were meant to be tallied in sequence, with the different clusters of notches within each column recording the days between the phases of the moon in several consecutive lunar cycles. From the Institut Royal des Sciences Naturelles de Belgique, Brussels.

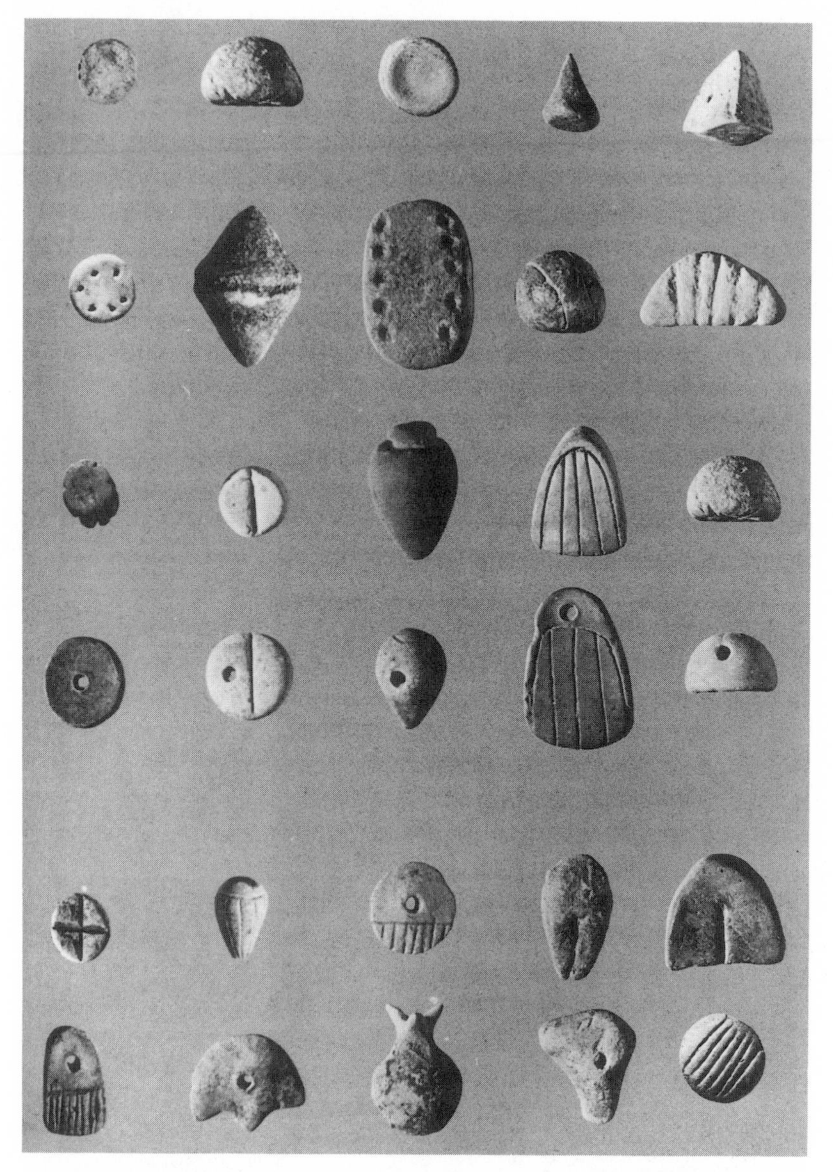

PLATE 2. *Clay symbols for record keeping.* Found on the site of the ancient city of Susa (in modern Iran), these symbols date from around 3300 B.C., just before the invention of writing. The pieces in the first row represent some basic shapes: sphere, half sphere, disk, cone, and tetrahedron. Various impressions and incisions render the pieces in the second row more complex. Perforations distinguish those in the fourth row from those in the third. The last two rows represent further variations in shape and markings. From the Musée du Louvre, Paris.

PLATE 3. *A proto-cuneiform tablet.* Dating from around 3000 B.C., this tablet combines pictographs made with a pointed stylus and circular impressions made with a round stylus. Some of the pictographs may already have phonetic values, and the circular impressions depict numbers. Neatly divided into columns and rows, the tablet illustrates the principle of emblem slotting, with each entry combining a name and a number. From the Department of Western Asiatic Antiquities, British Museum, London.

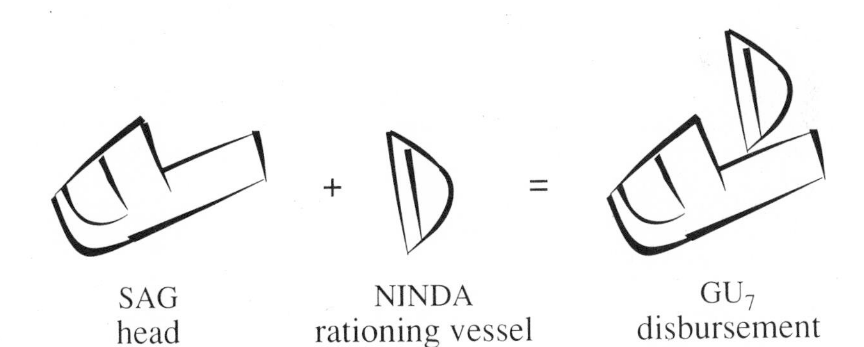

PLATE 4. *New words from old.* Proto-cuneiform scribes created the sign for "disbursement" by combining the signs for "head" and "rationing vessel." In this reconstruction, the signs are written horizontally rather than in the original vertical orientation of proto-cuneiform script, an innovation that parallels the growing abstraction of the script as phonetization became more prevalent (also see plate 6). From Hans J. Nissen, Peter Damerow, and Robert K. Englund, *Archaic Bookkeeping: Early Writing and Techniques of Economic Administration in the Ancient Near East,* trans. Paul Larsen (Chicago: University of Chicago Press, 1993), p. 15.

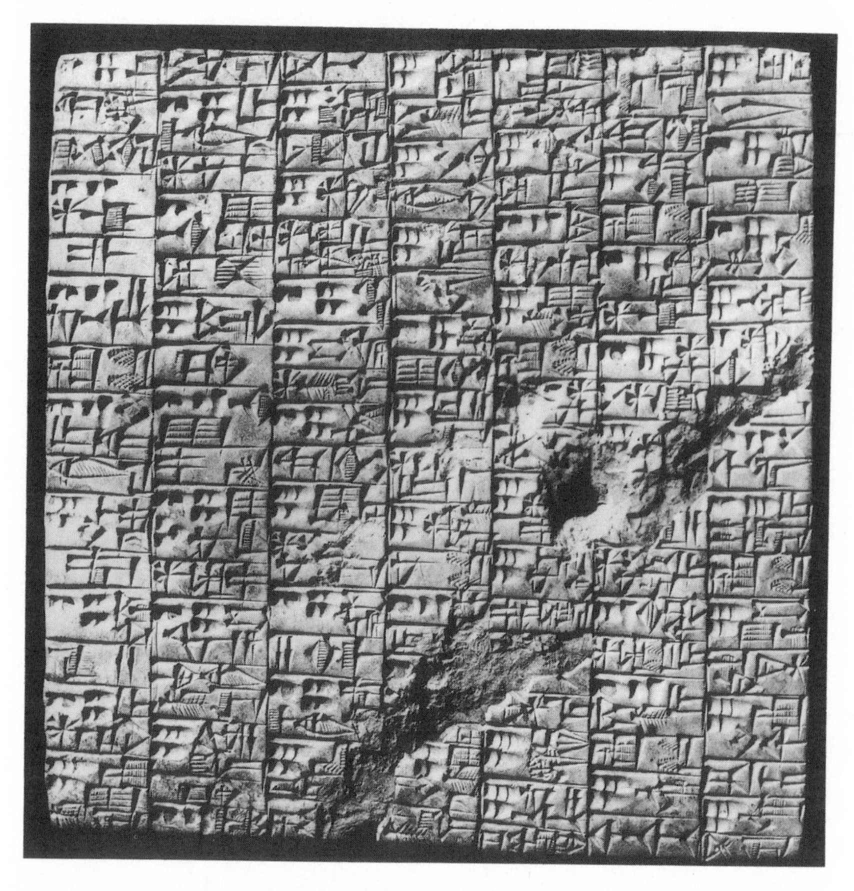

PLATE 5. *A cuneiform tablet.* Dating from around 2600 B.C., this tablet records the allotment of fields, registering the size of the field and the person to whom it was allotted. Instead of using a pointed stylus to make curved drawings, such as the bowl in plate 4, scribes now used a triangular-tipped stylus to render once curved signs with a series of elongated punch marks, thick at one end and thin at the other. Scribes could extend the thin end of the impression, also known as the "tail," to the desired length by keeping the stylus impressed while drawing it across the surface of the clay. From the Vorderasiatisches Museum, Berlin.

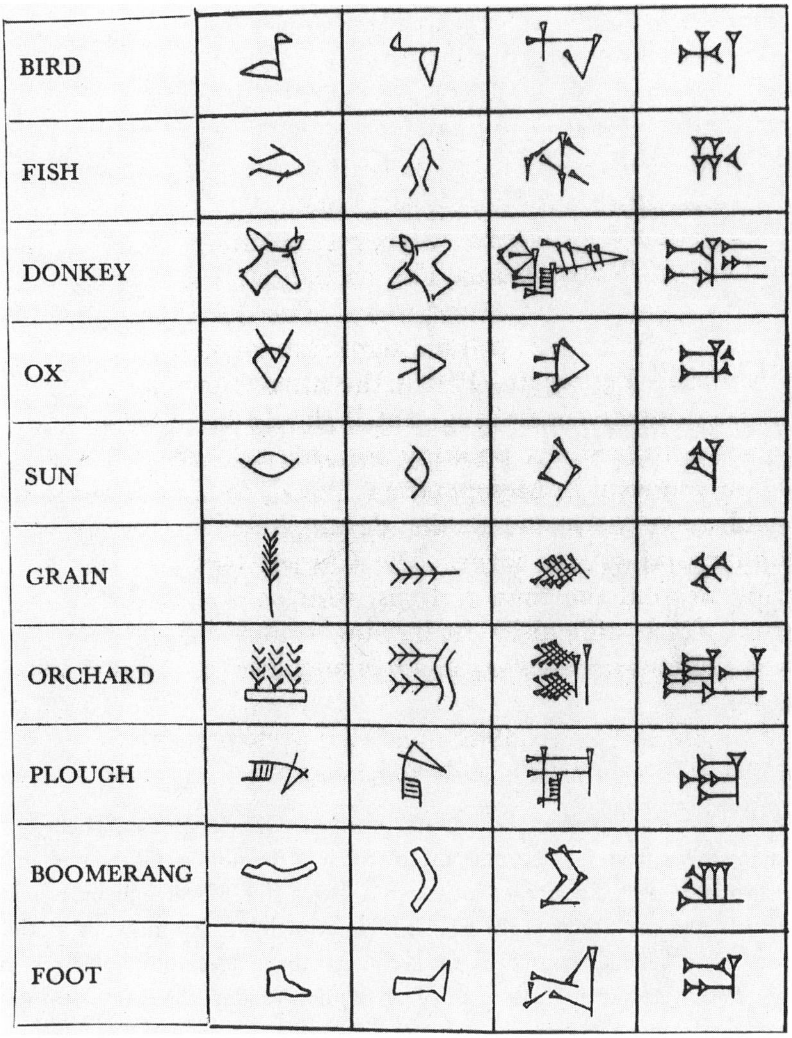

BIRD

FISH

DONKEY

OX

SUN

GRAIN

ORCHARD

PLOUGH

BOOMERANG

FOOT

PLATE 6. *The evolution of ten cuneiform signs.* From left to right, the columns illustrate the proto-cuneiform origin of these signs, their rotation from the vertical to the horizontal axis, and their initial and later rendition by means of a triangular-tipped stylus. From the Oriental Institute, University of Chicago.

CHAPTER THREE

Alphabetic Literacy and the Science of Classification

Wisdom is knowledge about certain causes and principles.

—ARISTOTLE, *Metaphysics*

The Effects of Alphabetic Literacy

In P. D. Eastman's perennial children's favorite *Are You My Mother?* a baby bird, having fallen from its nest, goes off in search of its mother. Of the many objects of its repeated query—a cat, a hen, a dog, a cow, a boat, a plane, a steam shovel—the dog's response is especially noteworthy: "I am not your mother. I am a dog." Strictly construed, the second sentence does not follow from the first. Yet an unspoken, classificatory assumption renders the entire statement crystal clear, even for a child.

Its meaning would be as evident to an ancient Mesopotamian child as a modern American one. We can assert this with confidence because the Mesopotamians must have been capable of making fundamental distinctions, and then some, in order to found and sustain a civilization. Human beings unable to differentiate between dogs and birds would not have thrived so visibly on this earth.

But (to let our imaginations roam further) the Mesopotamian equivalent of Eastman's book would have taken a very different turn. Instead of being about distinctions, it would be about things (dogs and cats; hens and crows; cows, sheep, and goats; ants and bees) arrayed by groups (pets, birds, livestock, insects) enumerable under the heading 'mothers'. Even though it might manifest many of the distinctions illustrated in Eastman's book, it would not encourage children to reflect upon them but rather to engage in a naming game. Between listing and classifying, between the Mesopotamians and us, lies that intellectual watershed we call "philosophy."

In the popular imagination, philosophy is a forbidding subject, so diffuse and ethereal (shall we say "abstract"?) as to appeal solely to pointy-headed intellectuals. Or so the caricature goes. At heart, though, philosophy is simply about our connections to the world, our ways of making sense of it. More often than not these function efficaciously, otherwise we would not be successful denizens of this planet, and to a large extent they are probably automatic, hard-wired by the process of biological evolution. Philosophy emerges when these ways of understanding become the subject of systematic, conscious reflection.

In this sense, Eastman's classic reads truly as a philosophical primer. It encourages children to reflect upon the differences between birds, cats, hens, dogs, cows, boats, planes, and steam shovels and, implicitly, to distinguish them further as animate and inanimate things. At some point a child, having learned to call various animate things by their specific names—'bird', 'cat', 'hen', 'dog', 'cow'—might briefly wonder how all these different things could also be known by a single name, 'animals'. But any such wonderment would quickly pass, to be replaced by the unquestioned assurance that the names we use correspond to the things of the world. This assumption makes us all, young and old alike, children of the Greeks. In asking precisely this kind of question, the Greeks first reflected systematically upon the relation between words and things, giving rise to philosophy.

Like any rich, fascinating subject, the origins of philosophy are, as Freud would say, "overdetermined." From the perspective of religion, the belief in immanent gods, dwelling within rather than beyond the cosmos, impelled the Greeks to examine the natural order of the world. From the perspective of culture, the Homeric poems instilled a powerful play element, which inclined Greeks to contend with each other in demonstration of intellectual as well as physical prowess. From the perspective of politics, the rise of Greek democracy accentuated the intellectual effects of this agonistic disposition, fostering a competitive environment of free expression and open inquiry. From the perspective of history, the long and ruinous Peloponnesian War encouraged the replacement of partisan, Homeric values with universal ones based on reason. And from our perspective on the technology of writing, alphabetic literacy broke the pictographic tie between words and things, forcing the Greeks to reflect systematically upon their connections to the world.

This kind of reflection would have been nonsensical in Mesopotamia, with its self-evident connections between words and things. So far we have referred by convention to pictographic signs as "words," but they did not really behave as elements of language in the strict sense. Rather, they depicted—or, better yet, were—the essence of things. Even as these signs evolved into syllables, and Mesopotamian writing began to pattern itself upon speech, they still retained something of their thinglike quality. Only the alphabet could break this tie completely, yielding signs that comprised words. Arbitrary symbols devoid of pictorial content, the letters of the alphabet depict sounds, not things. In alphabetic writing, meaning inheres not in the signs themselves but in the syntactic relations between words used in sentences. Language thus stands revealed visually as a system of meaning distinct from the reality it describes, engendering a disjunction between words and things.

The Greeks clarified their connections to the world through a hierarchy of terms mirroring the observable order of nature. Their remarkable success in this endeavor has led us to take for granted their classifying approach and the natural order it reflects. The animate beings in Eastman's story comprise two classes, 'bird' and 'mammal', which belong to the family, 'vertebrate', of the kingdom, 'animal'. We know an entity in this hierarchy, not by decoding its name, but by locating it vertically and horizontally in relation to its neighbors, by understanding its position in the natural order.

This form of knowing precludes endless naming. 'Dog' is not a category designed to facilitate the exhaustive naming of creatures grouped by arbitrary criteria (size, color, or the social status of their owners). Rather, it is a class with fixed characteristics, definable by means of its position in a hierarchy of classes. In knowing the class 'dog', we really know much more, for we are situated in the hierarchy such that we can ascend or descend at will, from class to class, without having to name all the members of each class. We can even skip classes—we can, so to speak, go to the head of the class—asking questions not about different kinds of being but about 'being' itself.

To the extent that it encourages the latter question, hierarchical thinking entails what Aristotle meant by *philosophia,* "love of wisdom." It circumscribes the endless naming of things by creating new kinds of information, in which abstractions from the flow of experience (say, the classes 'dog' and 'cat') themselves become the subject of further abstraction (yielding the class 'mammal'). The process of naming is delimited at each level of abstraction, which determines the field of inquiry. And these fields in turn are arranged hierarchically, in a process that reaches logical closure when one ascends to the ultimate causes and principles of existence.

We shall now turn to the advent of the alphabet, which culminated the primeval information revolution. By breaking the pictographic tie between words and things, the alphabet encouraged the fullest realization of the classifying possibilities of language, resulting in the emergence of a full-blown science of classification. This science would dominate the Western imagination for more than two millennia, until the seventeenth century, when modern numeracy would provide a new technology for informing experience.

The Nature and Origin of the Alphabet

The alphabet created the separation between words and things by rendering spoken language visible, by taking isolable sounds of speech and correlating them with abstract signs. We produce sounds of speech in two ways: (1) by forcing air through the vocal cords, causing them to vibrate; and (2) by stopping, restricting, or shaping this flow of air, chiefly by means of the tongue and the lips. Although the range of sounds produced by these means is vast, conventionally most languages utilize only about forty of them. Sounds created by a continuous flow of air controlled chiefly by the shape of the mouth have become known, in the wake of alphabetic literacy, as vowels; sounds created by stopping or restricting the flow of air at some point in the mouth have become known as consonants.

As the term implies, 'consonants' are sounded *with* vowels, having no voice of their own. For example, when asked to pronounce the consonant signified by the letter 'b', we say *bē*—the vowel gives voice to the consonant. The basic unit of pronunciation, the syllable consists of either a free standing vowel or of one or more consonants combined with a vowel. In any given language, the many different consonant-vowel combinations will produce hundreds of syllables.

In theory, syllabic writing models itself upon speech by providing a sign for each syllable used in a language. Because of the difficulty in remembering signs for hundreds of syllables, they naturally tend, by means of a "principle of economy," to reduce to a more manageable number. In Mesopotamia, the principle of economy encouraged the use of syllabic signs designating vowels but not consonants, such as one sign for the syllables *ga, ka,* and *qa.* In order to interpret the sign properly, the reader must infer which consonant should be paired with the vowel from both a knowledge of the language and the context of the communication. Despite being "syllabic," therefore, the signs do not actually specify pronunciation. Instead of having specific phonetic values, they serve as mnemonic devices for someone who already knows the pronunciation of the language and can thus choose from the range of possible values.

The alphabet is revolutionary because it overturns the previous relationship between writing and language. Rather than being a mnemonic device for recalling a language one already knows, it is a means of transcribing the sounds of language in general, regardless of whether we know their meaning. Because most languages limit the number of sounds to no more than about forty, the alphabet can (with slight modifications) be used to transcribe virtually any language. It accomplishes this feat by giving visual representation to the two classes of sounds formed by the human voice, vowels and consonants.

The real breakthrough here lies in the representation of consonants, which are unpronounceable by themselves. In other words, the alphabet reduces the basic unit of pronunciation, the syllable, to its components, one of which exists only as an abstraction. By symbolizing these components as "letters," the alphabet represents an aural phenomenon visually. Writing thus ceases to be a semiotic means of encoding information—whether about different kinds of livestock, different words beginning with the same syllable, different sorts of omens, different types of mothers—and becomes instead an efficient means of transcribing speech.

Its efficiency makes the alphabet a far more effective medium for the conveyance of information than any previous form of writing. In syllabic writing, the principle of economy entails ambiguity, for someone must always fill in the missing phonetic values from a knowledge of the language and the context of the communication. The ambiguity here affects both the form and the content of statements recorded in syllabic writing. The form of these statements tends to be repetitive, utilizing a limited vocabulary in order to minimize ambiguity, while the content tends to be dogmatic, favoring authoritative pronouncements, again in order to minimize ambiguity. (Recall the curiously flat quality of the revolutionary statement that inaugurates the Sumerian King List: "When the kingship was lowered from heaven, the kingship was in Eridu.") By contrast, the alphabet captures the nuances of speech, enabling one to convey much more information by means of writing.

The Greek alphabet derives from West Semitic scripts, which themselves stem from Egyptian writing (in distinction to the East Semitic scripts of Mesopotamia). Like its Mesopotamian counterpart, Egyptian writing is "word-syllabic," using about six hundred object signs and about one hundred syllabic signs. Whereas Mesopotamians economized on syllabic signs chiefly by indicating vowels but not consonants, Egyptians used exactly the opposite means, adopting one sign for all syllables with the same consonant or consonants. They did so because their language (like all Semitic tongues) is inflected, with the declension of nouns and conjugation of verbs being accomplished chiefly by retaining consonants while varying vowels—as, for example, in the English conjugation "sing, sang, sung." Were English written without vowels, readers encountering 'sng' would infer from the sentence's context that they were dealing with an instance of singing, along with the appropriate tense of the verb.

The Egyptian emphasis on consonants instead of vowels contributed significantly to the development of the Greek alphabet. Scholarly consensus still holds that the Greeks derived their alphabet from the West Semitic script of the Phoenicians, despite recent speculation about the influence of Aramaic (another West Semitic script) and Proto-Canaanite (a precursor of Phoenician script). Phoenician writing in particular represents a striking advance in economy over Egyptian writing. Whereas the Egyptians utilized about one hundred syllabic signs—of which about twenty indicate an initial consonant plus any vowel and about eighty indicate two consonants plus one or more vowels—the Phoenicians utilized only twenty-two signs, each indicating an initial consonant plus any vowel. For example, the syllables *ka, ke, ki, ko,* and *ku* were all represented by the sign 'k'. Other West Semitic scripts, slightly less economical, utilized up to thirty signs.

Most scholars believe that each consonantal sign derived from a pictograph that had ceased to represent an object, coming instead to represent the initial sound of the word for that object. According to this view, the pictograph for the Semitic word for bull (*'alf*) eventually came to represent the initial sound of that word (from which we get the letter *a*), and the pictograph for the word "house" (*bet*) came to depict its initial sound (from which we get the letter *b*). In the process of radically reducing the number of consonantal signs, the connection between West Semitic scripts and their pictographic origins became increasingly attenuated.

Those who study West Semitic scripts sometimes misleadingly term them "consonantal alphabets." In the above example, however, 'k' is technically not a consonant but a sign designating a set of consonants, for which the reader must supply the vowels. Words written without vowels are not readily pronounceable, as we can see in the following example: "Wrds wrttn wtht vwls r nt rdl prnncbl." This example illustrates how we can decode such writing if we know the language, even while the script itself does not clearly indicate pronunciation.

The Greeks adapted this syllabic form of writing to the pronunciation of their own language, using Phoenician signs with no equivalent sounds in Greek to stand for vowels. On this account, they are sometimes credited with creating the alphabet by means of adding vowel signs. This claim is a little misleading, for other scripts utilized vowel signs, though in a less systematic and consistent way. The achievement of the Greeks was to give visible expression to pure consonants by systematically representing vowels. The West Semitic scripts obviously laid the foundation for this accomplishment by creating a manageable number of consonantal sets. But the Greeks crossed a revolutionary divide when they distinguished between consonants and their attendant vowels, thus correlating visible signs with what were, for all practical purposes, aural abstractions.

We can only speculate about what enabled the Greeks to make this leap. Recall their love of hexameter verse, whose rhythmic qualities they enlisted in the service of social memory. Their exposure to Phoenician writing probably coincided with the fullest flowering of this oral culture. Both bard and audience celebrated the poetic tradition by means of formulas, which devolved upon aural as well as visual abstractions. Perhaps the Greeks, already accustomed to a high degree of aural abstraction, simply extended this process by analyzing the sounds of speech into their constituent parts.

Consensus holds that the Greeks developed the alphabet in the course of trade with the Phoenicians and, consequently, that the alphabet originally served accounting purposes. Our previous examination of the relationship between writing and accounting would seem to support this interpretation. But the standard story scarcely explains the revolutionary nature of the alphabet. If only accounting needs motivated the Greeks, surely an unmodified Phoenician syllabary would have sufficed to record a limited range of transactions. No doubt, the alphabet was devised in the course of trade, but probably to facilitate memory of speech—particularly the ever-present hexameters of epic poetry, snatches of which every Greek trader carried in his head—rather than business transactions.

The earliest examples of alphabetic writing in Greece corroborate this assertion. All broadly literary in nature, they consist of actual hexameters and what are probably hexameter fragments. This fact itself surprises, given the prevalence of accounting records and lists in the early stages of other forms of writing. Also, the earliest Greek inscriptions date from about 730 B.C., around the time that most scholars assume Homer flourished. We might even speculate that the genius of the Homeric poems spurred creation of this new, highly nuanced means of transcribing speech.

The adaptation of Phoenician signs for the transcription of spoken Greek marks the final divorce between writing and its pictographic origins. The Greek alphabet is twice removed from these origins, first because the signs of its West

Semitic progenitor had already begun to lose their pictographic associations and, second, because these signs were necessarily arbitrary for the Greeks, who borrowed them wholesale from a foreign people. The Greeks perceived the words formed with these arbitrary signs as representing reality only indirectly, through syntactic relations between words in sentences. These relations emerged all the more starkly given the alphabet's efficiency at representing speech. Henceforth, words would exist purely as words, as signifiers of things. We shall see below how efforts to understand the world would hinge not upon the decoding of words but rather upon their definition, a task that seeks to order knowledge by clarifying the relationship between the signifier and what it signifies.

Hesiod: From Genealogy to "Genos"

Narrative is the enemy of classification: Story telling recounts a temporal sequence of actions diachronically, while classifying describes atemporal, synchronic relations between entities. If the Greeks developed the alphabet to record hexameter verse, perhaps even the very epics attributed to Homer, what enabled them to break the constraints of narrative and found the science of classification? This question becomes all the more perplexing when we consider how ideally the alphabet preserved narrative. And yet alphabetic literacy ultimately realized the linguistic propensity to classify more fully than any previous form of writing, bursting the very narrative structure it had served so well.

An examination of the works attributed to Hesiod, an archaic poet second in importance only to Homer, reveals how the classifying impulse emerged from within the narrative tradition. In the *Theogony,* Hesiod creates a genealogy of the gods, a narrative that subordinates action to order. In the *Works and Days,* he extends this early classificatory effort, articulating a conception of *genos* that goes beyond the notion of "birth" to encompass that of "type." Too often we overlook Hesiod's importance because, as the intellectual descendants of the Greeks, we take his efforts for granted, without appreciating the hard-fought struggle between the narrative and classifying impulses.

Most scholars agree that Hesiod flourished about a generation or two after Homer, around 680 B.C., and that of the two works generally attributed to him, the *Theogony* predates the *Works and Days.* Although Hesiod's poems use many of the same figures and formulas as his predecessor, he hails from the world of the peasant farmer rather than heroic warrior, and he exploits poetic traditions native to his region of Boeotia, in addition to those of the Ionian epic.

One of these traditions is genealogical, concerned less with heroic acts (although these crop up) than with generational relations between individuals. Common to many cultures both oral and literate (recall the genealogies in the Book of Genesis), this tradition celebrates the fame of individuals, families, and cities, chiefly by relating them to heroes and gods. It appears on a small scale in the Homeric poems, which contain numerous genealogies of various heroes. Always brief, however, these rarely mention more than five or six ancestors. On this account, we may infer the tradition was less developed in Homer's region of Greece than in Hesiod's, where bards customarily sang genealogical poems in honor of their patrons.

A relative lack of narrative imagery—the scaffolding of oral memory—necessarily limited the length of these songs. By contrast, literacy enabled Hesiod to compose a genealogical poem of scope and grandeur. In the *Theogony*, he integrates many traditional stories, creating a majestic cosmology that encompasses the origins and nature of the gods and of the universe they oversee. These stories derive from a wide range of sources. Besides the anthropocentric gods of Homer's aristocratic warrior culture—such as Zeus and Hera—one finds the natural deities of an earlier agrarian culture—Gaia, Ouranos, Night, Chaos, and Cronos, among others. Stories about these gods circulated orally, but Hesiod may have had access to written versions as well, which he could study and compare as he integrated them into his all embracing account.

In devising his genealogy, Hesiod necessarily had to extract information about the gods from its original context. In some cases, this context was already genealogical, but not always. Much of his information comes from the Homeric poems, where immortals act no differently than mortals and participate fully in the human drama. Hesiod had to distill from these and other action narratives a more static picture of the gods, describing their nature, origins, and attributes.

The genealogy he created still narrates, although primarily begettings and births and only secondarily heroic deeds. Indeed, the *Theogony* presents in many places a list of who begot whom, unrelieved by heroic flourishes. As noted above, this kind of genealogical narrative is not an artifact of literacy, being characteristic of many oral cultures. But literacy enabled Hesiod to use it in an entirely new way, not to flatter a patron with a short song about his divine ancestry but rather to organize a multitude of diverse myths and legends in an extended poetic record. In an undertaking of this length, genealogical narrative became an instrument of order.

Subsequent poets extended Hesiod's genealogy of the gods into the mortal realm. The most famous of these works, the *Catalog of Women*, picks up where the *Theogony* leaves off. Attributed to Hesiod in antiquity, this work was actually compiled in the sixth century B.C. from sources dating back an additional two centuries. Although only a few fragments of it survive, philological reconstruction enables us to infer that this genealogy, which traces descendants from the female line (as was customary in some regions of Greece), integrated a vast amount of historical and geographical as well as social information. Indeed, it represents something of an encyclopedia of Greek culture, in which genealogy links together the heroes of all the great song cycles, including not only the Homeric epics but also those about Thebes and Hercules.

The genealogies of Hesiod and his followers constitute a Greek form of list making reminiscent of the Mesopotamian one. In both cases, literacy has begun to realize the classifying potential of language, exciting a desire for order. For the Greeks, though, this desire arises within a tradition of poetic narrative that had been transcribed by means of the alphabet. Within this tradition, genealogy provides the ideal classificatory tool, for it narrates a sequence of actions. It thus sustains the tradition while, at the same time, subjecting it to a hierarchical ordering that clarifies the nature of various figures. When gods are considered, genealogy becomes a means of understanding the cosmos, of establishing its eternal and immutable order; when mortals are considered, it becomes an encyclopedic framework for historical and geographical as well as social information. Genealogy, therefore, tends toward a synchronic view of the world, without realizing it entirely. Ultimately, it remains a narrative of actions, but one in which births supersede deeds and action serves the requirements of order.

In the *Theogony,* Hesiod classifies without transcending the limits of narrative; in the *Works and Days,* he takes the first tentative steps beyond narrative. Interpreting his poem poses difficulties, with its diverse and sometimes only partially developed themes, but in general we can describe it as an analysis of the human condition. Whereas the *Theogony* concerns the origins of Zeus's rulership over the cosmos in general, the *Works and Days* concerns the nature of his rule over mankind in particular. It essays a description of moral life, not of heroic action. In this work, Hesiod's attempt at analysis conflicts with his narrative instinct, revealing a literary mind whose opposing objectives struggled in constant tension. His ultimate failure to realize the goal of nonnarrative analysis should not blind us to the significance of his having made the attempt.

Hesiod writes of the human condition from bitter experience, addressing the poem to his brother, Perses, who had defrauded him of his rightful inheritance. This act prefigures one of the two types (*gene*) of strife, unhealthy discord as opposed to healthy competition. Hesiod describes the origins of these two types genealogically with reference to the gods mentioned at the beginning of the *Theogony,* but he seems to be expanding the notion of *genos* here beyond that of birth and lineage to encompass that of a type or class of actions. Indeed, he goes on to describe how Zeus rewards those who engage in the good type of strife while punishing those who engage in the bad.

In Homer's warrior culture, strife appeared always a dangerous force, for heroic self-assertion could go too far, as Achilles demonstrates, threatening friend as well as foe. Whereas Homer had drawn no clear line between healthy competition and ruinous discord, Hesiod explicitly split the heroic notion of strife in two, creating good and bad strife, healthy and unhealthy competition. This move may reflect Hesiod's more agrarian background, where strife entailed not only dangerous contentions between individuals but also fruitful ones between man and nature. Yet he could not have realized the possibility of a distinction between the two until the alphabet undermined the power of narrative, wrenching the *topic* of strife loose from *stories* about strife.

The alphabet accomplished this separation by accurately rendering the oral tradition in written form, so that one could compare different instances of strife and derive generalizations about it. Still, Hesiod's analytical tendency in the *Works and Days* conflicts with a narrative one, and he only goes so far in devising his general categories. Stories crosscut the poem everywhere. Among others, we find the famous accounts of Prometheus, Pandora, and the five ages of man. These stories illustrate Hesiod's notion of the human condition in a rather clumsy fashion, interrupting his would-be analysis with lengthy narratives that take on a life of their own. Clearly, Hesiod had difficulty engaging in atemporal description without falling back upon accounts of narrative action.

The same tension appears again as Hesiod proceeds from these famous stories into a description of farm life, an example of the healthy competition to which his brother Perses should aspire. This description, too, is storied, though in places it becomes little more than a list of proverbs serving to highlight types of behavior:

> Be a friend to your friend, and come to him
> who comes to you.
> Give to him who gives; do not give to him
> who does not give.
> We give to the generous man; none gives to him
> who is stingy.
> Give is a good girl, but Grab is a bad one.

The poem ends with an enumeration of the lucky and unlucky days for certain activities on the farm. As a collection of proverbial wisdom, it displays many characteristics of a list, even though the list making arises in the context of stories about farm life.

Given its mixture of literary motifs—the address to Perses, the description of good versus bad strife, the stories about various gods and ages of man, the description of farm life—the *Works and Days* lacks the coherence of the *Theogony*. This may in part be due to the corruption of the surviving text of the poem, but it also derives in part from the weakening of the narrative tradition—no single story line suffices to describe the human condition. Instead of an overall narrative, we find an overarching theme, the distinction between good and bad strife. Whereas the *Iliad* relates particular instances of strife, the *Works and Days* stands as a proto-analysis of strife in general. Hesiod thus brings us to the verge of classifying, a way of thinking that seeks to establish a category and analyze its various characteristics. Although Hesiod's habit of story telling ultimately undercuts his own hierarchical analysis, we have moved well beyond the additive, ever-widening inclusiveness of Mesopotamian list making.

Plato and the Art of Classification

In Platonic philosophy, narrative clearly yields to analysis. Lest we overemphasize novelty in Plato, we must begin by noting that much remains of the old oral habits of mind, which coexist with the new literate ones. First and foremost, philosophy for Plato is dialogue, an oral activity. True, the Platonic dialogues take written form, but Plato never lost his distrust of this invention. In the *Phaedrus* he warns that writing, as an external form of reminding, could weaken the exercise of memory, by which knowledge is internalized. And he laments that books are, if anything, more obdurate than the people who write them, for written words "seem to talk to you as if they were intelligent, but if you ask them anything about what they say, from a desire to be instructed, they go on telling you just the same thing forever." These concerns reflect oral habits of mind, habits that also leave their traces in the narrative framework of the dialogues, which always take place between historical personages in specific settings.

Despite this framework, the content of the dialogues concerns primarily the definition of atemporal states of being, not temporal and sequential actions. These definitions comprise Plato's famous "ideas" or "forms." From the perspective of our inquiry, the forms explicate the relations between words and things, which had become problematic in the wake of alphabetic literacy. And these relations are explicitly classificatory, for they reflect a hierarchy of terms that purports to mirror the structure of reality. Plato's, however, remains an *art* of classification, to be transformed into a *science* by Aristotle.

The *Parmenides* provides us with a simple example of the relation between words and things in Plato's theory of forms. Here Plato has his mouthpiece, the young Socrates, sketch the theory in response to an argument put forth by

Zeno. Parmenides, the greatest thinker of the older generation, then subjects Socrates' theory to numerous criticisms, in response to which Plato attempts (not always successfully) to refine the theory by working though various objections.

The dialogue begins as Socrates asks Zeno to confirm the opening hypothesis of an obscure argument Zeno had just finished presenting. Against the commonsense view of a kaleidoscopic reality, composed of many different bits and pieces, Zeno reaffirms (somewhat cryptically) that, "If things are many, they must be both like and unlike. But that is impossible; unlike things cannot be like, nor like things unlike." In an attempt to clarify this fuzzy thinking, Socrates posits the existence of the form "likeness itself" and its contrary, "unlikeness itself"; the things Zeno describes as "many," the unique entities of the world, then, partake of these forms. Those entities partaking of likeness are alike, those of unlikeness are unlike, and those partaking of both are both. Contrary to Zeno, things can be both like and unlike, but it is inconceivable, says Socrates, that the forms of "likeness" and "unlikeness" can be one and the same.

Socrates continues in a typically homespun way to explicate his meaning. As a human being, he is both one and many. He partakes of "unity" to the extent that he is an individual, distinguishable from other individuals; he partakes of "plurality" to the extent that he consists of many body parts. Thus, *he* is both one and many while the *forms* of unity and plurality remain distinct from one other.

In this discussion the theory of forms addresses a basic problem in naming. How can one thing (say, a human being) be referred to by several different terms ("one" and "many")? The problem would have been inconceivable in Mesopotamia, where the nature of the thing was inscribed in the word. By divorcing words from things, alphabetic literacy has raised questions about the appropriateness of one's terms. The theory of forms provides a tidy solution by explaining that the same thing can be called by different names to the extent that it partakes of different forms. Implicitly, the reverse also holds true, that different things can be called by the same name to the extent that they partake of the same form.

Plato establishes the relationship between words and things by means of "dialectic," which begins with a question about the meaning of a word. What is justice? Piety? Beauty? The question and answer method we call "Socratic" serves to gather together a range of possible definitions and then to distinguish between them, creating a hierarchy of terms by which one ascends, say, from various instances of justice to the idea of justice. Plato maintains that the realm of ideas revealed by dialectic is the true reality, eternal and immutable, from which our mutable world derives its characteristics. This position reflects the assumption that we grasp reality with the mind, rather than the senses; anything perceived clearly by the mind must be true. Ideas are thus not mere terms but real essences that inform the material world. For Plato, the realm of ideas explains the relationship between words and things, constituting the locus of all knowledge and meaning.

In the *Parmenides,* Plato subjects the theory of forms to penetrating criticism in order to reveal possible areas of weakness. One of these is that the theory makes more sense when applied to an understanding of intangible qualities—such as justice, piety, and beauty—than when applied to the tangible things of the material world. Parmenides demonstrates this problem to the young Socrates when he asks whether, in addition to the intangible forms, there is a form of man (apart from individual men), a form of fire, and a form of water. When Socrates affirms this, he falls into a trap; for Parmenides continues, asking whether there might also be forms for hair, mud, and dirt. At this point, Socrates hesitates between being consistent in the application of his theory and "tumbling in a bottomless pit of nonsense," in which the multiplication of forms threatens to duplicate the complexity of the material world.

The possibility of myriad forms for tangible things raises another question for Parmenides, concerning the nature of the relationship between a form and its manifestation in things. This problem—less apparent in discussions of justice, beauty, and piety—is cast in high relief when we consider the idea of man or the idea of the table. How is an object—*this* man or *that* table—informed by its idea? To the young Socrates' various attempts to field this question, by saying that the object either "partakes of" or "imitates" its idea, Parmenides raises a host of complex objections (the details of which need not concern us here). And if we do not know the nature of the relationship between ideas and their worldly manifestations, Parmenides continues, how can we know that ideas exist at all?

Though Plato offers no answer to these questions in the *Parmenides,* perhaps we can see the outlines of an answer in the *Sophist,* where he elaborates a taxonomy of forms extending downward from the "one" to the "many": "[The philosopher] discerns clearly *one* form everywhere extended throughout many, where each one lies apart, and *many* forms, different from one another, embraced from without by one form, and again *one* form connected in a unity through many wholes, and *many* forms, entirely marked off apart. That means knowing how to distinguish, kind by kind, in what ways the several kinds can or cannot combine."

In this explanation, the "one" is analogous to a genus, which is differentiated into the "many," its species and subspecies. The genus embodies all the characteristics of its member species and subspecies, which derive their essences from its essence. The real world of ideas is thus comprised of forms that extend from a kind of super-genus embodying everything—described variously by Plato as "Unity," "the One," "Beauty," and "the Good"—to a multitude of individual forms so specific they cannot be further differentiated. These have become known as the *infimae species,* the lowest species.

This hierarchy orders a multiplicity of forms, with even hair, mud, and dirt conceivably having their appropriate place. The multiplicity of forms is not only ordered but also delimited, for the lowest species mark the boundary between the true reality of ideas and the material world of the senses. Even if the hierarchy were to include hair, mud, and dirt, it could not be elaborated endlessly, dissolving into a mirror of the material world. The boundary is, furthermore, almost imperceptible, minimizing the separation between the two. We can more easily imagine how ideas inform corporeal things when the lowest species are almost (but not quite) as extensive as the entities of the material world.

For Plato, however, the fact remains that a separation between the two worlds must necessarily exist, and the exact way in which the real world of ideas informs the material one of objects remains unclear. Though he thought his theory the most reasonable explanation of reality, Plato had no means of convincing a determined skeptic, for nothing in the material world can prove the existence of the higher, true reality. Given its lack of a clinching argument, we may regard his hierarchy of forms as an *art* of classification, the distinctive product of his creative genius.

Plato might not be distressed to hear his philosophy labeled an art rather than a science. The latter term, from the Latin *scientia,* originally denoted exact, logically demonstrable knowledge. To the extent that such knowledge depends upon a written medium in which terms have fixed meanings, Plato disparaged it. Late in life, in his *Seventh Letter,* he belittled the writing of philosophy, as if all the dialogues he had composed were mere bagatelles, secondary to his true mission of teaching by personal example, in face-to-face communication with a select group of students.

Despite having originally articulated the theory of forms to explain the relation between words and things, Plato discourses in the *Seventh Letter* upon the instability of language, which invalidates the writing of philosophy: "Names, I maintain, are in no case stable. Nothing prevents the things that are now called round from being called straight and the straight round, and those who have transposed the names and use them in the opposite way will find them no less stable than they are now. The same thing for that matter is true of a description, since it consists of nouns and of verbal expressions, so that in a description there is nowhere any sure ground that is sure enough." Given the instability of language, Plato wrote dialogues not in order to provide the reader with logical proofs but rather to inspire him to pursue the *activity* of philosophy, which takes place in face-to-face conversation with others. Ideally, this activity overcomes the instability of language by fostering a meeting of minds, "when, suddenly, like a blaze kindled by a leaping spark, [knowledge] is generated in the soul and at once becomes self-sustaining." Furthermore, the blaze of knowledge can be kindled only in a select few, whose minds are truly receptive and, hence, capable of being nurtured patiently through the personal example of a dedicated teacher.

Knowledge is so utterly removed from the written word that anyone who commits his most serious concerns to writing is, to paraphrase Plato, witless. He may have chosen not to write a comprehensive account of his theory of forms, addressing all logical objections, because logic was ultimately a function of face-to-face communication. Only through dialogue could the mind leap the gulf separating the real world of ideas from the material world of objects. In leaping this gulf, the mind seizes whole the truths that are only partially illuminated by language, igniting a spark of knowledge that is truly self-sustaining.

Aristotle and the Science of Classification

As is well known, Plato's attitude toward knowledge and its relationship to language changes dramatically with Aristotle. The *Categories,* traditionally the first of his compiled works, begins by considering the three kinds of relations between things and the words used to express them. Things are named "equivocally" when they have only the name in common, as, for example, when a man and a portrait of the man are called by the same name. Things are named "derivatively" when one name stems from another, as, for example, when we derive the word "heroism" from "hero." And, most important, things are named "univocally" when they not only bear the same name, but that name means the same thing in all cases, as, for example, when we refer to both a man and an ox as "animals."

This discussion reveals an intellectual world markedly different from Plato's. In the first place, Aristotle has replaced the dialogue form, Plato's simulacrum of philosophical activity, with that form of logical analysis we nowadays associate with the term 'philosophy'. With this analysis, Aristotle assumes that words can be used univocally, that they have clear and unique referents in the world. Moreover, he bases this assumption upon another, that the clear relation between words and things derives from and reflects a hierarchical ordering of phenomena—an ox and a man are both classed as animals. Aristotle has supplanted Plato's ultimate distrust of words with a fundamental assurance that "language follows thought and thought follows things."

Whence this assurance? We cannot give a definitive answer, but some likely ones have been suggested. Son of a physician, from whom he may have learned to trust his powers of observation, Aristotle pioneered the study of biology, a form of inquiry grounded in the careful observation of living things. His faith in the possibility of univocal naming may well reflect the intellectual bias of a naturalist, for whom the senses provide accurate information about the order of life, an order everywhere apparent, in which each class of things can be clearly designated by a name.

To this explanation we should add another, that Aristotle's assumptions also reflect the influence of his teacher. After all, he studied for twenty years with Plato, from whom he learned three principles in particular: that there are forms, that these exist in a hierarchy, and that they are the true objects of knowledge. Generally speaking, Aristotle has simply taken these forms and brought them down to earth. Common sense led him to assert that, in the terrestrial world, "form" is inseparable from "matter" and that, therefore, knowledge must necessarily begin with entities as they exist materially. For him to assume otherwise seemed illogical.

In contrast to Plato, then, Aristotle regarded the material world as the true reality. He devoted much of his philosophy to describing (1) how we come to know this reality by drawing away from it, such that we can isolate and identify the entities that comprise it; (2) how these abstractions can be named univocally; (3) how the definitions of names mirror the natural order of the world; and (4) what the assumptions are upon which this order is based. Otherwise stated, Aristotle's philosophy stands as the culmination of the first information age, for it seeks to define and systematize the mental objects that had been wrought by literacy.

According to Aristotle, all knowledge begins with our sensory experience of the material world. This experience constitutes a continuum, perceived as a seamless flow of sensory stimuli. (Recall the bard's audience swaying to his song.) We are afloat in this sea of experience, which in and of itself offers no vantage point for knowledge. Nevertheless, knowledge will come of it.

Our minds operate upon this continuum, fixing our attention on particular aspects of it. These aspects then become fixed, extracted from the flux of the world. Aristotle designates these "mental extracts" with the Greek word *hóros,* "that which is limited," which aptly expresses his notion of things having been removed from the unlimited and undefined flux of the world. The Greek word originated from the expression for boundary marker, from whose Latin equivalent, *terminus,* we derive our word 'term'. A term is an expression by which we "define" or "determine" a thing, in the Latin sense of setting limits to it. A term is thus a logical concept, establishing the *lógos* of a thing, its "language," demonstrating with exactitude "what is said of it."

This language, for Aristotle, derives from the thing itself. One might object that the mental extract is the product not of the thing but of a mental process we call intuition and, as such, has no connection to reality. Aristotle would agree that the extract is made by our intuition. But he would insist that intuition adds nothing to our sensory perception. Rather, it only takes something out of perception, and in so doing defines and delimits it. The extract thus derived is not a mere phantasm but a mental object distilled from a physical one. Words, therefore, have a real connection with the things they refer to, making it possible to speak meaningfully about the world without appealing to a separate, ideal realm or reality.

Strictly speaking, the extracts of intuition are prelogical, but because they arise from things, they already reflect the order of reality. For example, when we see a dog, our intuition fixes itself upon, and thereby fixes, certain features, say for the sake of illustration, that they are 'four-legged', 'furry', and 'barking'. We extract these features because they appear to define or delimit what it is to be a dog, whereas other features—such as those pertaining to size, color, and markings—do not. Much like P. D. Eastman's young audience, we know what a dog is by automatically contrasting it with other extracts—bird, cat, hen, cow. At the level of prelogical extracts, therefore, we are already beginning to sort out objects, in effect, classifying them.

This intuitive process, however, is not yet one of "knowing" things but only of "perceiving" them. According to Aristotle, we introduce knowledge when we make the classifying procedure explicit. We use words to formulate definitions from our intuitions, definitions whose truth can be tested. Thus we might assert on the basis of our intuitive extracts that 'A collie is a dog'. In making this statement, we have gone beyond the intuitive process by which we derived the term 'dog'. Now, instead of extracting the term from individual entities, we are reversing the process and applying it to them. In other words, we are asserting its generality, claiming that it constitutes a class of things. Our intuitive extract becomes a full-fledged logical concept when we give it a precise definition in order to test its generality. In the case of our hypothetical example, we might say, 'A dog is a four-legged, furry, barking animal'. Strictly speaking, at this point 'dog' becomes a term, in the sense of a logical concept.

The definition analyzes the intuitive extract into its parts, which establish the boundaries of the thing under consideration. How did we derive these boundaries? As mentioned above, the intuitive extract 'dog' emerges from an implicit contrast with other extracts, from which 'dog' is differentiated. The first step in making a definition is to name this larger group. Thus, we might begin by saying, 'A dog is a four-legged animal'. This definition, however, does not get us very far, so we now move on to differentiate 'dog' from the other members of this group. We might do so by dividing the group into smooth-skinned and furry animals—a dog is thus 'a four-legged, furry animal'. This definition, however, still encompasses extracts other than 'dog', so we review them one by one, seeking their distinctive features. The process of enumeration is potentially open-ended, though it concludes when we think we have arrived at the practical certainty that a dog is a four-legged, furry, barking animal.

The boundaries of the extract reflect a hierarchy of genus and species. In the case of our hypothetical example, we have 'animals', of which some are 'four-legged', of which some are 'furry', of which some 'bark'. Through the analysis of the extract into its parts, the definition articulates and orders the classes we have grasped intuitively.

What knowledge have we gained by transforming our intuitive extract into a concept? On the face of it, very little, for the concept adds nothing to the extract. But in defining the extract, we have at the very least analyzed it explicitly into its parts, which otherwise would have remained submerged. And these parts define or delimit the extract with reference to a hierarchy of genus and species. Once we understand the concept's position in this hierarchy, we gain knowledge of it beyond what is implicit in the extract. For example, when we define a dog as an animal, all the properties of the genus 'animal' become true of 'dog', properties that may not be readily apparent in the extract. Likewise, when we break the genus down into the class 'four-legged', an even more specific set of properties may be predicated of 'dog' that also are not necessarily apparent in the extract. The definition establishes the concept's position in a hierarchy, after which we are flooded with knowledge from above.

Our hypothetical definition of a dog as 'a four-legged, furry, barking animal' reveals a problem underlying Aristotle's science of classification. According to this definition, seals might also be termed 'dogs'. How can we be sure that we have classified properly?

If, given confusion between dogs and seals, we are unsure whether to define dogs as 'furry, barking animals', we can simply ascend to the next highest level —in this case, 'four-legged'—and begin to work our way down again. Confusion, however, may still result—in the case of our example, dogs and frogs might end up being classed together, to say nothing of birds and men. If so, we can ascend further up the hierarchy, to the genus 'animal', and differentiate it according to other, observable criteria. For example, Aristotle distinguishes between egg-laying animals and those that birth their young, thereby setting dogs apart from frogs, and men from birds. As we work our way down again, 'four-legged' may yet come into play, at a lower level of differentiation. For Aristotle the hierarchical order of categories guarantees that we will find some purchase point for definition, a point of sufficient generality that we can know it with certainty, from which we can proceed to differentiate its constituent classes.

By explicitly transforming intuitive extracts into concepts, we create information. Of course, Aristotle does not use this term, nor does he associate this logical process with literacy. But as the term *lógos* implies, logical processes concern the precise use of words: Such precision flows as a natural consequence from literacy in general. And, more to the point, the mental objects he calls concepts are functional counterparts to the abstractions of alphabetic literacy in particular. Concepts are not (as with the products of Mesopotamian writing) coextensive with the things of the world; rather, they are a further step removed from the world—in our parlance, they convey information about it. Precisely on this account, the nature and status of conceptual information needs to be analyzed; its connection to the world needs to be established. In so doing, we draw away from this information; we begin to abstract from our abstractions, creating new kinds of information at each remove. This process reveals a taxonomic hierarchy that, for Aristotle, mirrors the order of nature.

Aristotelian philosophy is grounded in common sense. It simply serves to define material entities in a process that naturally reveals the taxonomic order of the world. There is, however, an unsettling difficulty with this naturalistic vision. When we ascend from the species 'man' to the class 'mammal', we exclude all those properties "specific" to man. Likewise, when we ascend from 'mammal' to the family 'vertebrate', we exclude everything specific to mammals, and everything specific to vertebrates when we ascend to the kingdom 'animal'. The higher we ascend the more generic or general—that is, empty—the classes become, until we arrive at the apex of the hierarchy, which contains . . . nothing!

Aristotle avoids this absurdity by ultimately basing his logical process upon metaphysical principles. Recall that knowledge begins with our sensory experience of material entities, from which we extract concepts. What enables us to do this? So far, we have seen how we define an extract with implicit reference to other extracts (a dog cannot be a bird's mother because it is a dog). In this process we grasp what Aristotle calls the "substantial form" or "essence" of an entity. For example, from a material entity (Socrates), we extract a concept ('man'), which is defined as such by a substantial form or essence ("the human soul") common to all members of the same species.

Aristotle takes pains to explain his idea of substantial form in a way that distinguishes his naturalistic philosophy from Plato's ethereal one. As a naturalist, Aristotle was convinced that all inquiry must begin with the things of this world, which common sense indicates are instances of formed matter. The term 'matter' is dense with meaning and crosses the Styx into a confusing philosophical netherworld. Suffice it to say that for Aristotle 'matter' does not denote (as it does for us moderns) the physical stuff of existence. Rather, it is at one level a logical posit, the necessary complement to form. And at a deeper level, it is a metaphysical principle underlying the taxonomic order of nature. From the existence of this order, one can infer the theoretical existence of pure potentiality, matter in the abstract that gets differentiated into myriad substantial forms.

The higher we ascend in the taxonomy, the farther we move from pure potentiality and the closer we come to pure actuality, to "being" in the abstract. For Aristotle, 'being' does not denote the ultimate genus, the Platonic source of all forms. Rather, like matter, it is a metaphysical principle inherent in the entire hierarchy of logical concepts, without which reality, the concrete instances of formed matter, could neither exist nor be understood. Reality thus lies in distended array between the poles of pure potentiality and pure actuality.

Although Aristotle imports metaphysical principles from outside his logical system to guarantee its truth, we may nonetheless regard it as constituting a science of classification. After all, the term 'metaphysics' literally means the book in his corpus following the one about physics, a book dealing not with ethereal matters but with the fundamental nature of reality. And, as observation reveals, reality is fundamentally governed by the actualization of potential—acorns grow into oaks and infants into adults. Although these principles lie outside the logical system, they are nonetheless self-evident to anyone who reflects upon the order of nature.

We have come a long way from the oral world, in which cultural continuity entailed the immersion of consciousness in the flow of ritual narrative. Under the pressure of accounting needs, people began to draw away from this "enclave of contrived speech," to see how speech itself functioned by classifying. As an invention, the genius of writing lies in its ability to exploit this classificatory potential, creating a new mode of communication independent of speech. And with writing was born information, as mental objects were abstracted from the flow of experience.

Writing redoubles the linguistic propensity to classify by making it possible to list things. Once initiated, list making acquires a momentum of its own, as the distinctions within and between lists reveal the possibility of other distinctions, and so on. The play element in culture naturally intensifies this tendency, eventually entailing lists of lists. Writing thus naturally encourages a sense of hierarchical order.

In Mesopotamia, however, the impulse to classify was constrained by the pictographic origins of writing. The mental objects abstracted from experience were perceived not as words but as embodiments of the essence of reality, to which the things of the world conformed. Although writing in Mesopotamia gradually began to pattern itself upon spoken language in order to enhance its capacity to carry information, it never lost the thinglike quality inherent in its pictographic origins. Knowing was thus limited to an endless naming that militated against the natural tendency to abstract from one's abstractions.

Not until the advent of the Greek alphabet was writing utterly divorced from its pictographic origins, permitting the full emergence of hierarchical thinking. In this new form of literacy, the objects abstracted from the flow of experience came to be perceived as *mental* objects, as words whose relation to things needed to be secured. It should now come as no surprise that the Greeks were able to secure this relation by taxonomic means, which is simply the fullest realization of the classifying possibilities of language.

The classifying view of the world culminates with Aristotelian philosophy. His taxonomy purports to reveal the reasons underlying the order of things, reasons explaining why a thing has the qualities that characterize it as a particular kind of thing. These "reasons why" are known as "causes," and thus, for Aristotle, exact knowledge—the product of logical definition—devolves upon a knowledge of causes. As one contemplates causation in general, one ascends from exact knowledge—what the Greeks termed *epistēmē* and the Romans *scientia*—to *sophia* or *sapientia*, "wisdom."

Wisdom, then, sits atop taxonomical science, giving closure to the process of abstraction. Indeed, it embodies the ultimate abstraction, so far removed from the things of the world that the pursuit of it, philosophy (the love of wisdom), exists for its own sake. Or so Aristotle claims. From our perspective, however, the knowledge of ultimate causes and principles carries a different implication. It obviates our having to name all the classes of things, a prospect almost as endless as the naming of the things themselves. Philosophy thus functions to circumscribe the information born of literacy.

Whereas the Mesopotamians had delighted in enumerating words precisely because their pictographic signs embodied the principles and causes of things, the Greeks with their alphabetic signs could take little comfort in this activity. Instead, they had to find principles and causes not in the signifiers themselves but in the order of the things signified. The absence of a divinely inscribed reality forced them to deduce from this order metaphysical principles guaranteeing the reality of their abstractions. Our own unspoken faith in the relation between words and things derives from this reflexive move, making us all children of the Greeks, descendants of the age of wisdom.

PART II

The Modern Age of Numeracy

CHAPTER FOUR

Printing and the Rupture of Classification

Scribbling seems to be a sort of symptom of an unruly age.
—MICHEL DE MONTAIGNE, *Essays*

A Surfeit of Books

In 1571 Michel de Montaigne retired from the parlement of Bordeaux to his country estate, intending to live the remainder of his life, "now more than half run out," in sagelike tranquility. He was only thirty-eight years old, young by our standards, but he preferred to see the glass as half empty. And with good reason. His beloved friend, Étienne de La Boétie, had been dead for eight years, the victim of a sudden and violent dysentery. His father, Pierre, had died three years earlier in slow agony of kidney stones, a malady that would eventually afflict Montaigne. A younger brother had recently died of a freak accident while playing tennis. And death was all around him in the guise of civil war between French Protestants and Catholics, which had been sputtering and flaring for some nine years now and would soon erupt in the Saint Bartholomew's Day Massacre. All this occasioned morbid thoughts.

Death would be the test of his life, or so thought the young retiree. To contemplate his end, he refurbished a tower of his chateau as a library retreat from mundane distractions. Here he hoped to attain wisdom, a tranquility and constancy of spirit impervious to pain and death. The goal was that of *philosophia,* for the knowledge of ultimate causes and principles served moral as well as ontological ends. Aristotle's successors, the Stoic and Epicurean moral philosophers of Greek and Roman antiquity, had advocated cleansing the mind of worldly concerns and ascending the hierarchy of being, to live in harmony with the principle of Reason that pervaded the cosmos and underlay the order of nature. Montaigne's library would enable him to follow this age-old path toward wisdom by affording examples of other lives to contemplate, lives that would spur him to greater philosophical detachment, lives revealed chiefly in works of history, poetry, and moral philosophy.

But it was not to be. Many of the books in his library had been bequeathed to him by La Boétie on his deathbed. Surrounded by this ever-present reminder of his loss, Montaigne could not but grieve for his friend. A sage should inure himself to the pain of grief, and so Montaigne sought solace in accounts of how others had dealt with loss. But these accounts offered no sure guidance. Ancient, medieval, and modern; historical, poetical, and philosophical—they all contradicted each other. So, too, did accounts relevant to the other moral and practical subjects he contemplated in the solitude of his library. Disparaging these unproductive meditations as "chimeras and fantastic monsters," he began recording them in order to "make my mind ashamed of itself," an effort that gradually evolved into a twenty-year intellectual journal, the *Essays*.

The size of Montaigne's library, which would eventually total a thousand volumes, fed the unruliness of his mind. Barely a hundred years earlier, before the spread of printing, a comparable library of manuscript books would have been worth a king's ransom and would have been difficult to amass in places other than a court or cultural center. By the end of the sixteenth century, printing had enabled a minor nobleman living in a provincial backwater to acquire what would have once been a princely collection. This well-stocked instrument of his retirement spurred Montaigne to question the relevance of the ancient ideal of wisdom to the modern world. Instead of living according to the model of sages past, he learned to live according to his own model, which he fashioned for himself as he contemplated the record of human diversity. Of course, the quality of Montaigne's thought cannot simply be reduced to the quantity of books available to him; but it is hard to imagine that the *Essays*, or the mind of its author, could have existed before the full flowering of printing.

Our own immersion in typographic culture conceals from us the impact of the brute abundance of books upon those who experienced the first age of printing. Consider for a moment the ruminations of one modern scholar: "A man born in 1453, the year of the fall of Constantinople, could look back upon a lifetime in which about eight million books had been printed, more perhaps than all the scribes of Europe had produced since Constantine founded his city in A.D. 330." The estimate of scribal book production is, to be sure, speculative, and the estimate of printed book production is, if anything, conservative. But the general point lies beyond dispute. And a man born in 1453 would have lived chiefly during the age of "incunabula" (from the Latin for "swaddling clothes"), denoting books from the infancy of printing. This period ends by convention at 1500, after which the number of presses and the output of published works increased voluminously.

Once firmly rooted, printing transformed the intellectual landscape. It enabled Europeans to secure their heritage of classical texts against the threat of loss or corruption, thus helping to make the Renaissance revival of antiquity permanent and contributing to our modern attitude toward the past. Similarly, printing secured advances in a wide range of fields, from anatomy to zoology, encouraging the development of modern science and medicine. And printing was the fulcrum of the Reformation, enabling religious dissent to inspire a mass movement. In short, it was an epoch-making invention.

But at its most basic level, printing gave individuals access to a previously unimaginable number of books, overloading them with diverse and contradictory information. Much of this information came from traditional medieval works that had gone into print early on: theologies, cosmologies, natural philosophies, romances. Although these publications added but little to the preexisting stock of information, they brought that entire cache out into the open, exposing the medieval worldview to a much more thorough and critical analysis. This accomplishment alone would have had revolutionary consequences, but its effects were redoubled by the recovery and restoration of classical texts entirely unknown in the Middle Ages or known only partially and imperfectly. And, in addition to the growing contrast between the ancient and the medieval, modernity itself began to take shape as recent scientific theories, religious doctrines, and geographical discoveries quickly found their way into print.

This avalanche of books and ideas engendered not intellectual advancement but confusion, undercutting the traditional, classificatory means of information management. The erosion itself was the unintended by-product of the Renaissance fascination with the power of rhetoric. Greek alphabetic literacy had sired the rhetorical as well as philosophical traditions, and Aristotle had classified both, relegating each to its appropriate place in the structure of knowledge. Whereas philosophy used logic to prove truths, rhetoric used eloquence to persuade of probabilities; and just as logic revealed the hierarchy of concepts and classes, so too rhetoric displayed a hierarchy of topics and arguments. Taken together, the two realms disposed all human experience between them, but they shared an uneasy equilibrium, with the proponents of one occasionally trying to subordinate the other.

Struggles between the two intensified during the Renaissance, when the topics of rhetoric came to rival and eventually to surpass the concepts of philosophy as the chief means of information management. These rhetorical means, while still taxonomic, organized information less by logical than moral principles. Accepted moral categories reflected a limited range of human values and beliefs, narrowly Christian, classical, and Western European. Printing filled these categories with examples of a much wider range of human experience: from Eastern Europe, Turkey, China, and the New World; from non-Christian peoples both ancient and modern; from popular as well as learned culture. Many of these examples contradicted the accepted moral categories, gradually weakening them. And as these categories lost their normative value, they ceased to be effective means of ordering information.

Growing uncertainty about the status of traditional moral norms undermined the whole classifying view of the world, as people began to question the very capstone ideal of wisdom. Doubts about the ability to know the order of the world catalyzed a crucial change, away from taxonomic forms of information storage based on natural language and toward new ones based on a symbolic language of analytical abstraction. Mathematics promised a new vision of order for both the natural and the moral worlds, where confusion was resolved by jettisoning whatever could not be known with certainty. The spread of printing thus marks a divide nearly as great as that created by the genesis of information, a divide separating the classificatory view of the world characteristic of literacy from an analytical vision born of numeracy.

Medieval Antecedents

Information overload did not so much originate as culminate with printing, which intensified a problem whose intimations surfaced several centuries earlier. Scribal document production in the Middle Ages had already advanced to the point where it required the development of increasingly sophisticated systems for managing information. The earliest of these systems simply involved the physical arrangement of information on a page; later systems applied more and more elaborate taxonomic principles to the problem of information management. Ultimately, advances in medieval book production engendered the comprehensive organization of all knowledge in a system that wedded Aristotelian philosophy to a Christian worldview. If even a relatively small increase in the number of books could call forth such elaborate taxonomic efforts, we can well imagine how much greater the impact of printing would be on the classificatory tradition.

The foundations for medieval advances in scribal production were laid in late antiquity, with the creation of the "codex" or book format for manuscripts and with the substitution of parchment for papyrus. The codex originated sometime in the first century A.D. and gradually supplanted the papyrus roll or *volumen*, the traditional form of ancient manuscript. Papyrus volumes were inherently difficult to consult, each consisting of a roll about nine or ten inches wide and about thirty feet long, containing roughly the same amount of text as a modern chapter. Imagine having to "scroll through" such a volume looking for a particular passage, a situation much like that of having to use a huge computer file without the benefit of search and retrieve functions. These difficulties encouraged the reader to memorize much of what he or she read, referring to the text only when absolutely necessary. By contrast, the codex made it possible to consult manuscripts more easily, simply by flipping to the desired page. Facilitating random access to information, the codex had the potential to transform the manuscript from a cumbersome mnemonic aid to a readily accessible information storehouse.

This potential emerged after parchment supplanted papyrus, making books more widely available. The use of parchment arose in the same period as the codex, largely because papyrus was extremely expensive and available only from limited sources, chiefly Egypt. Although the use of parchment may have done little to offset the expense of writing in antiquity, it encouraged book production in the Middle Ages, when skins were commonly employed for this purpose. The finest parchment, vellum, was made from calf skin and remained expensive, reserved for special books like the Bible. But skins of lesser quality also came into use for a wide range of books and documents, providing a readily available, relatively inexpensive, and extremely durable writing material.

Medieval book production expanded noticeably in the twelfth century. Until then it had taken place in monastic *scriptoria*, where monks wrote manuscripts in laborious, formal book hands, chiefly for use within their monasteries. With the rise of universities lay stationers also became involved in book production. For the most part, they relied on a "putting-out" system, in which they divided an approved copy of a manuscript into sections and farmed them out to local scribes for duplication. Book production may also have occurred in secular *scriptoria*, with assembled scribes copying from dictation (although some scholars now question the existence of such manufactories). But it definitely did occur under similar circumstances in university lecture halls, where students copied texts dictated by their professors.

The adoption of cursive scripts and smaller book formats reflects the increased demand for books. Cursive scripts were quicker and easier to write than the more formal, monastic book hands. They also took up less space, resulting in the production of smaller, easier-to-use books. The products of the monastic *scriptoria* tended to be large, bulky volumes, designed to be displayed publicly on lecterns and read aloud to an audience. In contrast, lay stationers produced works chiefly for individual, private use, where the expense and inconvenience of a large format were unnecessary.

Increased scribal book production dating from the twelfth century has led some scholars to claim that a medieval "book revolution" occurred three hundred years before printing. This claim, however, exaggerates the significance of medieval developments. Although book production grew and the range of works became more diverse, output still remained relatively small. Before printing, the most learned individuals owned only a few precious volumes.

Even this limited growth in book production fostered new means of organizing knowledge. The rise of universities stimulated the book trade and led to the establishment of centrally located libraries that were easier to consult than those in remote monasteries. Although the output of books remained small, access to them grew, and changes in book formats made them inherently easier to use. Individuals thus confronted the need to assimilate an increasing amount of diverse and sometimes contradictory information.

One of the most visually striking means of doing so was through the technique of glossing a text with commentaries. The term 'gloss' (from the Greek word for 'tongue') reflects the medieval practice of dictating text and commentary during university lectures. Students glossed their texts between the lines and in the margins, using a small cursive script to distinguish the commentary from the text. Some manuscripts, resold by university stationers, developed successive generations of glosses, each in a different hand. Stationers also produced preglossed texts, covered with the most famous commentaries (see plate 7).

Glosses were used especially in the study of the Bible and of Roman and canon law, where students and stationers could readily annotate the small number of key texts. These glosses incorporated the opinions of all the major commentators. The passage being explicated was written in red or marked by some other means, and the different sources of the commentary flowing around it were identified by symbols, figures, or letters. For example, different Biblical commentaries (say, those of Saint Augustine and Saint Jerome) could be identified and indexed in the margin of the page by different patterns of colored dots.

Organized in a highly visual and spatial manner, the glossed manuscript became a compendium of knowledge. The elaborate pictures, distinctive symbols, and ornate letters characteristic of illuminated manuscripts served as visual cues, mnemonic hooks aiding in the recall of information. Even the visual pattern formed by the juxtaposition of text and commentary on the page —each one had its own unique shape—served to imprint the contents of the page in memory (see plate 8). As the number of respected commentators grew, however, the glosses threatened to swamp the text in a morass of opinions.

Another means of organizing and managing information was the *summa,* or "summary," which—unlike the gloss—aimed at cutting through the diversity of interpretations to establish a core of truth. The progenitor of the *summa,* Peter Abelard's *Sic et non* (Yes and No), was composed in the early twelfth century. It illustrates a range of theological issues by gathering together groups of contrasting texts drawn from Scripture and the church fathers. Abelard wanted to provide the raw material for logical disputations, so that students could find their way through theology's "mass of words" in order to establish its central truths.

Whereas Abelard simply contrasted authorities on selected theological issues, subsequent thinkers sought to traverse the logical process of reconciling the authorities and resolving the issues, hence giving rise to the age of the *summa.* Around 1150 Peter Lombard composed one of the first and most popular of the *summae,* his *Libri quattuor sententiarum* (Four Books of Opinions). This work follows what became the standard pattern of the *summa,* beginning with a *quaestio,* a question or field of inquiry, then listing the contrasting pronouncements of Scripture, church councils, canon law, and the church fathers on the question, and finally offering a *sententia* or conclusion to the question. More a popularizer than an original thinker, Peter Lombard had simply intended to write a "textbook," as it were, of theology. This textbook itself, however, quickly assumed authoritative status and became the subject of commentaries that (ironically) served to gloss his *summa.* The phenomenal success of his work reflects the tremendous desire to synthesize the diverse authorities that had become more readily available in the twelfth century.

The synthesis that Abelard desired and Peter Lombard provided was based chiefly on fragments of Aristotelian logic preserved by Boethius in late antiquity. With the geographical expansion of European civilization in the twelfth and thirteenth centuries, the whole corpus of Aristotelian philosophy was recovered from Greek and Arabic sources. This recovery supplied theologians with more powerful tools for synthesizing and organizing an increasingly diverse array of opinions. The philosophical revolution wrought by the recovery of Aristotle culminated with the *summae* of Thomas Aquinas in the mid to late thirteenth century.

Aquinas fashioned the whole philosophical tradition that had come down to him (centering chiefly on Aristotle but also incorporating Plato and the neo-Platonists, as well as Arab and Jewish thinkers) into the unified, harmonious instrument of a truly Christian philosophy. In the greatest of his two *summae,* the unfinished *Summa theologica,* he endeavors to explain the principles of Christianity to the extent reason permits. His motivation for doing so, as he declares in the prologue, is to cut through the confusion created by too many books on the subject of Christian doctrine, books that bewilder the reader by raising a host of repetitious questions and arguments in no particular order.

The three parts of the *Summa theologica* bring classificatory order to this confusion. The first part begins at the beginning, with God, and moves on to consider His Creation in a systematic fashion, starting with the angels and culminating with man; the second part considers the nature of man in detail; and the third part examines the nature of Christ, the Savior of man. Within each part, the argument proceeds taxonomically, from general to specific. The analysis of man, for example, starts with a consideration of the ultimate end or purpose of human life. It then considers various aspects of human nature, such as the will by which man achieves his end and the passions that advance or retard his progress. This topic leads ultimately to a consideration of the principles guiding human action, as embodied in law, which is (in turn) analyzed according to the dictates of divine, natural, and human law.

Within this Aristotelian framework, the pattern of analysis follows the traditional form of a *summa,* which Aquinas elaborates in a series of logical steps that separate each question into its constituent parts. First he states the question or topic (concerning, for example, the existence of God). Then he breaks the topic down into a logical series of articles (whether the existence of God is self-evident, whether His existence can be demonstrated, whether He actually exists). Then he briefly states his position on each article, reviews the chief objections to his position, lays out the reasons and authorities supporting his position, and refutes each objection in turn. In this manner, he proceeds systematically, step by step, from article to article and, ultimately, from question to question. The intellectual edifice of the *Summa theologica* has been compared to a Gothic cathedral, both in the harmony of its finely wrought details and in the immensity of its structure.

Although the framework is Aristotelian, the content of Aquinas's philosophical system is fundamentally Christian, harmonizing diverse philosophical schools from the perspective of Church doctrine. The Aristotelian system of knowledge begins with the material things of this world and moves upward through a conceptual hierarchy of species and genera, each of which is more general than its predecessor. Were it not for metaphysical principles underlying the notions of matter and form, the apex of this hierarchy would be entirely empty of content. Aquinas, however, begins at the apex, with God, whom he treats as a creative as well as classificatory principle. In this way, he resolves some of the anomalies and fills some of the gaps in Aristotelian philosophy, harmonizing it with other schools of thought in order to elucidate Christian doctrine.

Culminating with Aquinas, the theological *summa* tendered a remarkably comprehensive means of organizing information within the framework of a Christian worldview. Although his synthesis would subsequently be called into question by those who either rejected the primacy of Aristotle or interpreted him differently, the *summa* format dominated the study of theology well into the seventeenth century. In fact, the practice of summarizing authorities and arranging them systematically also became commonplace in other subjects, such as the study of law, which like theology was marked by the proliferation of judgments and authorities in the Middle Ages.

The very success of the *summa* format serves to underscore a limitation inherent in the systematic arrangement of authorities and opinions, namely that it places excessive emphasis on logical procedures. One can easily be put off by the forbidding form of a *summa* that moves relentlessly from question to question, article to article, objection to objection—step by mincing step. One can also be put off by the kinds of seemingly ridiculous questions that arise by logical necessity in the systematic exploration of each topic. For example, in analyzing the nature of angels (disembodied, nonmaterial beings), we might logically consider whether an angel can be in several places at once, or whether several angels can be in the same place at the same time, say dancing on the head of a pin. Finally, one can be put off by a logical language whose very abstruseness grows along with the complexity of the questions under consideration. For later critics, such logic chopping transformed the whole of scholastic thinking into a cartoon, a caricature of serious thought.

Some of these objections to scholasticism arose in the Renaissance, from the fourteenth through the sixteenth centuries, when rhetoric came to rival philosophy as an instrument for understanding the world. The growing emphasis on the art of persuasion led to the adoption of rhetorical categories for organizing and managing information, in place of philosophical concepts and classes. Although still thoroughly classificatory, the rhetorical categories reflected a moral rather than logical hierarchy, one specifically concerned with understanding the human world, apart from the natural one. The resulting intellectual system, which we shall term "commonplace thought," preserves the taxonomic form of information storage but lacks its logical spine. With the spread of printing, commonplace thought would prove too fragile to sustain the weight of information generated by printing.

Rhetoric and the Rise of Commonplace Thought

Nearly as old as philosophy, the rhetorical tradition dates from the sophists of the fifth century B.C., who taught public speaking. Plato has succeeded in giving sophists a bad name, as seen in our notion of sophistry as deceptive argumentation. In a similar vein, we are accustomed to denigrating 'rhetoric' with the adjective 'mere', as though the effective use of words amounts to little more than hot air. This modern bias obscures the supreme importance of rhetoric in the world before mass media, when so much communication took place in face-to-face venues. Athenian citizens, for example, discussed community affairs informally in the public square, debated issues formally in periodic citizen assemblies, and represented themselves before large juries in law courts. In context, Plato's attacks on the sophists were hardly disinterested, for his greatest rival in Athens was the rhetorician Isocrates, with whom Plato competed for students.

Aristotle resolved the rivalry between philosophy and rhetoric by situating each in its proper category or realm. Philosophy concerns matters of truth, whereas rhetoric concerns those of probability, the gray areas of life where truth cannot be determined with certainty. When we stop to think about it, we live much of our lives in these gray areas, especially in matters of public policy and the law, where we constantly mediate between possibilities and strive to determine likelihoods. Aristotle offered a reasoned procedure for doing so by arguing on both sides of any given question—both for and against—and he also showed how to substantiate this procedure by means of generic types of arguments. These can be found in what he terms the *topoi,* topics or places containing various approaches to argumentation. Not surprisingly, he divided his "places" by genus and species: They are either "common," containing the kinds of strategies applicable to all arguments regardless of subject matter, or "specific," containing strategies appropriate only to specialized subjects.

The Romans upset the balance between the art of rhetoric and the science of logic, so expanding the notion of the "common" places that it eventually overshadowed the "specific" ones, constituting a means of organizing all knowledge, including philosophy. A people of simple peasant origin, the Romans never cared much for philosophizing, preferring to concern themselves with practical matters, mostly with questions about the gray areas of life. As they absorbed Greek culture, they found more use for its rhetorical than philosophical tradition. Indeed, of the schools of Greek philosophy, the Romans especially admired Academic skepticism, which asserted that one could not know truth but only weigh probabilities by arguing on both sides of a question.

Rome's greatest orator, most prominent Academic skeptic, and most influential proponent of Greek culture, Marcus Tullius Cicero subtly transformed the Aristotelian notion of the "common" topics. In addition to regarding them as the places where one could search for general kinds of arguments, he also treated them as if they comprised the particular stuff of argumentation—ideas and expressions that all audiences took for granted, to which a successful orator must appeal. In this sense they are akin to our notion of "commonplaces" as proverbs or maxims, except that we regard such expressions as well worn, if not trite, whereas Cicero employed them as basic categories of thought. Through them the orator accessed and applied the information necessary for decision making in a world where truth could not be ascertained.

The practice of reading in antiquity underlay and reinforced the Ciceronian notion of commonplaces. Readers naturally tended to distill norms of human conduct from their readings, regardless of whether an author had consciously intended to convey them. Through a process of concentrated reflection, readers imprinted such noteworthy passages upon memory by breaking them down into pieces of information (examples and snatches of phrase) keyed to visual icons, striking images—a dagger, an eagle—that triggered recollection of the passage's form and substance. They then stored these images in the appropriate mental categories or "places"—represented in the mind's eye as the columns of a temple, the diners at a banquet table, the objects in a room, the rooms in a house—whose spatial organization structured the entire mnemonic system. The more images associated with a passage, or the more places where an image could be stored, the greater the number of pathways for accessing the form and substance of the passage. By means of an art now long lost, the ancients built up memory visually and spatially, place by place, layer by layer, creating a flexible mnemonic network that enabled them to retain the contents of a papyrus roll without having to consult the actual text.

These "places" stored not "raw data" but the form and substance of ethical concepts, truths the society took for granted; hence, they were also known as "commonplaces." Readers shaped their character by furnishing their memory with these moral norms. The process of judgment involved sorting through the norms and adapting them selectively to the occasion at hand. For Cicero, the ideal orator combines wisdom with eloquence in such a way as to apply his accumulated stock of norms most appropriately and most fully to the occasion. Societal truths thus anchored the orator's skeptical mode of arguing both sides of a question, giving him the means to arbitrate between contrasting positions. Far from being trite expressions, commonplaces tender nothing less than the common ground of social discourse.

The Roman tendency to blur the Aristotelian distinction between common and specific places, fostering the overarching notion of commonplaces as societal truths, was accentuated by the practice of compiling *florilegia*. Forerunners of *Bartlett's Quotations*, these "bunches of flowers" consisted of proverbs, maxims, and eloquent expressions culled from great writers to ornament speeches and compositions. The compilation of such textbooks originated in antiquity and flourished in the Middle Ages, when it constituted yet another means of managing the information generated by increased scribal book production. The organization of these textbooks varies, but as a rule the literary flowers are bunched around moral themes, combining the places where one searched for generic arguments with the actual stuff of argumentation. And the themes themselves, arrayed hierarchically, serve to classify the moral world in the traditional taxonomic manner. With the rise of rhetoric in the Renaissance, these kinds of commonplace collections would evolve into encyclopedias for the organization of knowledge.

In the Middle Ages, rhetoric was subordinated to philosophy, with the *summa* providing the most systematic and thoroughgoing organization of knowledge. The mnemonic techniques of classical rhetoric continued to thrive in this period, as attested by the popularity of *florilegia*, but the commonplace norms of the well-furnished mind were subsumed under a system of divine truth elucidated by logic. With the rebirth of classical Latin eloquence, however, Renaissance humanists asserted the superiority of rhetoric over and against the philosophy of the scholastics. In so doing, they unintentionally weakened the logical framework supporting the taxonomic form of information storage.

The humanist movement originated in fourteenth-century Italy, not long after the coming of scholasticism to that region. Polemical battles between humanists and scholastics were not so much between old and new ways of thought as between contemporary ones locked in hot competition. In this contest the humanists seized the rhetorical high ground, lamenting the "dark age" that preceded their efforts to revive the light of eloquence, that is, to revive Latin in its classical purity, cleansed of what they called medieval "barbarisms." A practical attitude toward the world underlay this initiative, in which virtuous action counts for more than knowledge of virtue. Francesco Petrarca, foremost of the early humanists, expressed their position best when he declared: "The object of the will is to be good; that of the intellect is truth. It is better to will the good than to know the truth." Better to love virtue with the heart—to lust for it—than merely to understand it with the mind.

Classical Latin eloquence provided the most effective means of "willing the good." Again Petrarca: "Everyone who has become thoroughly familiar with our Latin authors knows that they stamp and drive deep into the heart the sharpest and most ardent stings of speech, by which the lazy are startled, the ailing are kindled, and the sleepy aroused, the sick healed, and the prostrate raised, and those who stick to the ground lifted up to the highest thoughts and to honest desire." Under this boundless faith in the power of Latin eloquence lies the conviction that persuasion outweighs reason in the practical world of human affairs, a world of gray areas amenable to rhetoric, not logic.

Although the primacy of philosophy long remained unshaken in the universities, humanists gradually established the primacy of rhetoric in secondary education. They did so by means of a curriculum, the *studia humanitatis,* which chiefly comprised grammar, rhetoric, history, poetry, and moral philosophy. These were the "studies worthy of human beings": grammar and rhetoric because they improve the capacity to communicate (whereas logic merely provides techniques of analysis); history, poetry, and moral philosophy because they elevate human thought to the level of action (whereas physics and metaphysics only concern knowledge of man and nature in the abstract). The progressive educational curriculum of its day, the *studia humanitatis* was well established in schools across Europe by the end of the sixteenth century, providing a cultural seal of approval for those who sought careers in both church and state.

This curriculum was inculcated by means of commonplace notebooks, the direct descendants of medieval *florilegia.* Students compiled these notebooks in the course of their readings in order to create a stock of ideas for their own speeches and compositions. In this way they furnished their minds and formed their characters with norms representing the common ground of social discourse. Desiderius Erasmus, the most influential humanist of the sixteenth century, recommended that students make their own notebooks, organizing them by topics paired with their opposites for easy recall, such as "piety" and "impiety," each of which was subdivided hierarchically. For example, he showed how to analyze the category "piety" into that toward God, the fatherland, the family, and so on, an arrangement descending from highest to lowest, from general to specific, in the classificatory manner.

No doubt a gap divided pedagogical theory and practice. Less-industrious students, say, might simply have annotated and cross-referenced a standard classical text like Virgil's *Aeneid,* noting in the margins similar ideas and examples encountered in other readings. After the spread of printing, one could even make use of published commonplace books in a wide variety of topical and alphabetical formats.

Whatever their diverse forms, commonplace books were all alike in that they comprised encyclopedias of classical culture. (We here use the term 'encyclopedia' not only in its modern sense, as a compendium of knowledge, but also in its traditional sense as a "circle of learning.") Commonplace books brought together all the knowledge one needed to know. This circle of learning was specifically classical because antiquity, especially classical Roman antiquity, represented the acme of cultural achievement. In the heights of their virtue and the depths of their vice, the ancients provided the most profound and enduring examples of action to imitate or to avoid. Moreover, classical authors provided the very model of eloquence, revealing and preserving the enormity of their cultural achievement.

The encyclopedic conceit of Renaissance commonplace books—that they encompass all one needs to know—distinguishes them from medieval *florilegia.* This conceit grew out of the conviction that moral action in the world counts for more than intellectual understanding. From this perspective, rhetoric no longer serves as the handmaiden of philosophy but supplants it. Incapable of intellectual certainty, the human mind can only mull over probabilities, the realm of moral action. As the art of applying the common stock of wisdom fully and appropriately to the occasion, rhetorical persuasion governs this realm. Ethics thus replaces logic at the center of the circle of learning, and the moral truths that had previously been illuminated by the divine light of reason now shine in the light of eloquence.

The Breakdown of Commonplace Thought

Our focus on notebooks and mnemonics may seem a strange way of epitomizing the Renaissance, for these offer (to our modern sensibility) mere pedagogical techniques. Yet they embody the ideals of the humanists, ideals rooted in antiquity and expressed in a curriculum, ideals that would shape the minds of those who bore the brunt of printing's impact. Like their medieval predecessors, people trained in the humanistic tradition fashioned their minds as information storehouses divided and subdivided into myriad places. Unlike their medieval predecessors, however, they assumed that each discrete category of stored information was significant in and of itself, without reference to a universal philosophical system that arrayed the categories in a logical hierarchy explaining the world. In other words, each piece of information exemplified a moral or societal norm, whose truth everyone took for granted. This assumption enabled the humanists to play at being skeptics, to examine an issue from both sides, secure in the knowledge that accepted norms would illuminate the side of greater likelihood.

This comforting assumption would be shattered during the course of the sixteenth century, which witnessed a religious reformation, the discovery of the New World, and the Copernican revolution, to cite but three momentous challenges to accepted beliefs. Such challenges helped weaken the social consensus that bound the system of commonplace thought. As they lost their normative vitality and viability, commonplaces ceased to be effective categories for managing the information subsumed under them. They became arbitrary placeholders, incapable of providing meaningful order to the profusion of information generated by printing.

Ironically, humanism unwittingly undermined its own intellectual system, opening the European mind to the epochal challenges it faced. The humanists aimed at reviving classical Latin eloquence and, along with it, the glory of classical culture. Attempts at restoring Latin to its ancient purity necessarily drew them into the historical study of language, into determining which words and expressions characterized classical Latin composition. Only in this way could they root out the medieval barbarisms that had crept into copies of Roman texts. Yet the study of language revealed that Latin literature had evolved through a succession of styles, which paralleled developments in Roman history. As knowledge of classical language, literature, and history grew, antiquity began to appear as historically distinctive, as unique and unrepeatable, calling into question the relevance of classical norms for the modern world.

Humanism thus forged the thin edge of the wedge of relativism that split the European mind from previously unquestioned truths. As modern Christians grew to appreciate the uniqueness of the ancient pagan world, they began to open themselves to the distinctiveness of other cultures, other religions, other views than their own. No sooner had the sense of historical and cultural relativism struck at the normative heart of commonplace thought than the spread of printing delivered the coup de grâce, overloading the moral categories of thought and discourse with so much diverse and contradictory information that, in effect, they ruptured.

In the sixteenth century, printing shed its "swaddling clothes" to reveal its productive potential. We opened this chapter with the estimate of 8 million books published during the period of incunabula (1450–1500), exceeding more than a millennium of scribal manuscript production. The estimate is overly conservative when compared to the calculations of Lucien Febvre and Henri-Jean Martin, Frenchmen who authored one of the seminal histories of printing (translated as *The Coming of the Book*). By their tally, the period witnessed the publication of between thirty and thirty-five thousand editions, of which some were small runs, numbering only a handful of copies, and some were relatively large, numbering 1,000 to 1,500 copies. Reckoning that the average edition ran five hundred copies at most, Febvre and Martin arrive at a figure of 15 to 20 million books published. And this just for the age of incunabula, when printing was barely established in Europe. By the end of the sixteenth century, a flourishing industry had produced 150,000 to 200,000 editions, larger ones averaging around a thousand copies, for a total production of some 150 to 200 million books!

To gauge the full impact of printing, we should consider not only the sheer number of books produced but also their variety. According to Febvre and Martin, the thirty to thirty-five thousand incunabula editions encompass ten to fifteen thousand different titles, a figure that fails to do full justice to the diversity of publications. In an age before critical editions of standard texts, different editions of the same text might include widely different material. Furthermore, different texts were often bound together and issued under the same title. An edition of Virgil's *Aeneid*, for example, might have another classical or poetic work appended to it.

Traditional works predominated during the first age of printing. Again according to Febvre and Martin, about 45 percent of incunabula editions are religious in nature (including the Bible, the church fathers, and scholastic theology), a little more than 30 percent are literary (whether classical, medieval, or modern), a little more than 10 percent are legal, and about 10 percent concern natural philosophy (what we would today term "scientific" works). Staples of medieval thought and culture form the vast bulk of publications in each category. At first glance, then, printing would seem to have confirmed established traditions—and it did. But in so doing, printing brought the full length and breadth of the medieval storehouse of information into the light of day, whereas before (given the relative paucity of manuscript books) it had been glimpsed only piecemeal. As medieval traditions became visible in their entirety, the view of the world they constituted became increasingly relativized, especially by contrast with the classical view.

During the sixteenth century, religious publications steadily gave ground to classical ones as a percentage of overall production. In fact, the sixteenth century is one of the great ages of classical scholarship, which helped restore the literary works of Greek and Roman antiquity to their original form in critical editions. Additionally, demand for vernacular translations of classical works grew apace, making the whole of the classical tradition widely available. These developments might simply have supplanted medieval norms with classical ones were it not for the relativizing effects of humanist scholarship, which had revealed the inimitable quality of classical culture. Further, toward the end of the century, accounts of new explorations and new approaches to science and medicine began supplementing traditional travel literature and natural philosophy, evoking the distinctiveness of modern as well as ancient and medieval culture.

Changes in book formats enhanced access to this increasingly diverse storehouse of information. The earliest printers had imitated the scribal practice of numbering the signatures, or collections of leaves, that comprised a book. In the course of the sixteenth century, this practice gave way to that of consecutive pagination. Likewise, other features of books that we now take for granted became regularized, such as the use of tables of contents, running heads, and indexes. Printing made these devices, employed only haphazardly by scribes, easy to accomplish and hence routine. Not only could readers find their way around a book more easily, but they could also more readily navigate collections of books by means of published catalogs and bibliographies, which proliferated in the sixteenth and seventeenth centuries.

The outpouring of printed books, as well as the newfound ease of accessing and using them, engendered information overload. People had available a much wider range of material than ever before. In sheer scope this material would have resisted assimilation, even if much of it had not been so diverse and contradictory. The common response to overload (now as well as then) is to shut down, to ignore information (about strange peoples, for example) that cannot easily be assimilated, and to disregard information (about religious and scientific matters) that contradicts accepted beliefs. But humanistic education had unwittingly opened a few minds to the effects of relativism. And information overload filled these minds, which had been fashioned according to the traditional system of places, with so much diverse and contradictory information that the places burst from within, necessitating that order be improvised.

The rupture of commonplace thought threatened the philosophical closure achieved by the ancient Greeks. Aristotle had placed wisdom, knowledge of the causes and principles of things, atop taxonomical science, obviating the need to name all the specific entities embraced by his hierarchy of knowledge. Although commonplace thought was not avowedly Aristotelian, it rested upon his classificatory assumptions, which had been applied to an understanding and ordering of the moral world. As the commonplaces at the heart of this intellectual system became increasingly relativized, people began to question the assumptions underlying not just the moral order but any kind of order. Without a sure means of classifying, they fell back upon the process of exhaustive naming in an effort to master information stripped of its traditional frame of reference.

A sure sign of information overload, lists are what we draw up when we have too much on our minds. The sixteenth century stands as one of the great ages of list making and related activities, as manifested in compilations of commonplaces, in lexical, herbal, and zoological lists, and in systems of "places" for the encyclopedic organization of knowledge and memory. Of course, not all list making signifies intellectual crisis (today's shopping list seldom becomes a *cri de coeur*), and in certain fields, like botany and zoology, which made dramatic advances with printing, lists are quite natural. But when list making spills over into unexpected areas, chiefly literary, we are justified in calling attention to it as a symptom of intellectual dislocation.

Monk, physician, and humanist, François Rabelais is easily the most notorious list maker of the sixteenth century. His excessive inventories of words, expressions, and things in *Gargantua and Pantagruel* display considerable ingenuity and serve a variety of purposes. On the most obvious level, they poke fun at traditional learning, as in the catalog of the Library of Saint Victor in book 2. Among the 150 or so titles are such gems of ersatz scholasticism as *The Codpiece of the Law, The Mustard-Pot of Tardy Penitence,* and *The Old Shoe of Humility.* On another level, his lists reflect the encyclopedic exuberance of a popular culture going into print: Witness the inventory in book 1 of Gargantua's childhood games (about 220 of them) and the compilation in book 3 of 160 slang expressions for Friar John's testicles (followed by another, equally long, for Panurge's privates). Rabelais delighted in immortalizing such popular pastimes and crass expressions.

On the deepest level, Rabelais's list making discloses a keen desire for order, especially apparent in the lists associated with eating and defecating. Typical of these is the outrageous list of Gargantua's ass-wipes in book 1, starting with a lady's velvet mask and climaxing with the neck of a well-downed goose. Lists like this bring a host of unlikely objects into contact with the human body, which Rabelais celebrates as the new touchstone of existence in a world where traditional norms have begun to break down. A compulsive quality pervades his celebration of the body and the sensual world. The lists are "Rabelaisian" not so much in their excessive coarseness as in their coarse excessiveness, in their ironical pretense to reorder the world exhaustively, a pretense that Rabelais mocks by their very length.

Montaigne's *Essays* reveal the same compulsion for order in a form of narrative list making integral to his mode of composition. Superficially, his writings resemble a traditional genre of commonplace literature that gathers examples from history, poetry, and moral philosophy around a central theme or moral. But the resemblance ends at the surface, for the array of examples really serves to highlight the diversity and complexity of the world, a conclusion Montaigne punctuates with extended lists of competing beliefs, bizarre customs, and unexpected actions. He intersperses these diverse examples with meandering descriptions of his personal habits and qualities. Finally, he binds the whole melange together as the "catalog" (*rolle*) and "record" (*registre*) of his ongoing musings, presented to the reader as if in a stream of consciousness.

The register of his mind provides him with a point of orientation amid the diversity and complexity of the world, a point that resolves itself, gaining focus as he essays or tests his mind against such diversity. In its own way this process rivals Rabelais's list making in its exhaustiveness. Montaigne proudly declares, "I have no more made my book than my book has made me—a book consubstantial with its author." In pursuit of this elusive consubstantiality, he multiplies his adjectives, compounds his clauses, and redoubles his parenthetical expressions, creating a "thick description" of the self that aims at completeness. Notwithstanding his literary grace and charm, which enables him to communicate with us more directly than any other sixteenth-century author, modern readers commonly experience a certain wooziness when perusing the *Essays,* a sense of intellectual claustrophobia evoked by the sheer density of Montaigne's self-portrayal.

Many writings of this period strike the modern mind as immensely cluttered. This quality appears conspicuously in Jean Bodin's *Method for the Easy Comprehension of History* (1566), a widely read guide to that body of literature. One modern reader describes this work as a "strange semi-ruinous mass," a heap of scholarly odds and ends—systems of note taking, climate and humor theories, astrology and numerology, and various chronological schemes, to name only a few. These schemes and theories serve to organize a body of literature vastly expanded by printing. They also serve to manage a conception of the past that had become filled to bursting with accounts of historical entities—peoples, states, religions, and cultures—each of which Bodin recognized as unique.

The widespread recognition of historical and cultural uniqueness, evident in works as diverse as Montaigne's *Essays* and Bodin's *Method,* underscores the breakdown of commonplace thought. Ancients differ from moderns; Brazilian cannibals differ from Europeans; Frenchmen differ from Italians. All these differences defy reduction to common categories. Montaigne put it best. Summing up a long list of contradictory philosophical opinions about one and the same subject, he declares sarcastically, "Now trust in your philosophy; boast that you have found the bean in the cake, when you consider the clatter of so many philosophical brains!"

The Skeptical Crisis

Commonplace norms had anchored in social consensus the skeptical mode of arguing both sides of a question. Secure in their moorings, the humanists had played at being skeptics, touting rhetoric over philosophy without really having to face the intellectual consequences of their position. Now, with the rupture of commonplace norms, the anchor was loosed. Some of those raised in the humanistic tradition, especially those sensitive to the effects of historical and cultural relativism, found themselves drifting toward a more radical skepticism that questioned the very possibility of establishing any truths for organizing and managing knowledge. Failure to resolve this skeptical crisis by humanistic means led eventually to the emergence of a new vision of order, one that abandoned the classifications of natural language, seeking greater certainty in a symbolic language of analytical abstraction.

The withering skepticism of the "Apology for Raymond Sebond," the massive central chapter of Montaigne's *Essays,* epitomizes the crisis ensuing from the breakdown of commonplace thought. Although most of the arguments voiced here are derivative, culled from classical authors, Montaigne forges them into a powerful weapon with which he hammers away at the presumptuous faith in human reason—that it makes man better than other animals, that it makes him happy, that it makes him good. Reason is merely "an instrument of lead and of wax, stretchable, pliable, and adaptable to all biases and all measures." Its instability derives from a soul bound to a body always in flux. As Montaigne wryly observes, "On an empty stomach I feel myself another man than after a meal." A mind thus incarcerated can have no sure grasp of the world.

Montaigne's assault on human presumption culminates in a critique of the senses that undermines the very foundation of Aristotelian rationalism. Through a variety of simple observations (that, for example, we perceive an oar to bend in water) Montaigne reveals the weakness and unreliability of the senses upon which reason depends. Reason fluctuates, then, not only with the condition of the body but also with its own inherent instability. This argument constitutes a direct assault upon the classificatory tradition of Aristotle, who assumed that our senses provide a clear window onto the world from which we can extract concepts. By emphasizing the weakness of the senses, Montaigne undermines confidence in any classification, whether the taxonomic hierarchies of Aristotelian natural philosophy or the commonplace systems of humanist moral philosophy.

Montaigne distilled his radical skepticism into the famous motto *Que sais-je?*—"What do I know?" Yet such was the nature of his genius that he could offer an answer to this seemingly insoluble question, on the basis of which he tendered a new means of organizing knowledge. His skepticism was not an end in itself but a way of searching for truth, of ceaselessly essaying his mind upon the diversity and variety of the world. In the course of these exercises, he willingly entertained the intellectual challenges most of his contemporaries preferred to ignore. In his famous essay "Of Cannibals," he gives a clear-eyed assessment of the inhabitants of the New World, whose practice of eating their enemies he judges more humane than the European treatment of captives. And in the "Apology," he takes a dispassionate look at the role of habit in religious faith, openly declaring, "We are Christians by the same title that we are Perigordians or Germans."

Montaigne's practice of essaying his mind provided him with a new means of organizing and managing information, one that mirrors the form but not the substance of commonplace thought. His essays are sufficiently like the commonplace literature of his day that contemporaries often mistook them as such. The most personal ones, given over almost wholly to self-portraiture, bear commonplace titles, like "Of the Useful and the Honorable," "Of Vanity," and "Of Experience." We can even read these compositions as collections of famous quotations and examples. Yet we can also see how Montaigne uses these borrowings not to illustrate accepted norms but to illuminate his self as he muses upon the flux of the human condition.

The self thus replaces societal truths at the vital center of commonplace thought. In a diverse and fluctuating world, the only thing Montaigne could know was the movement of his own mind as he contemplated that diversity. And commonplaces, distilled from his readings and stored in notebooks, could still provide him with occasions for contemplation. As such, they continued to organize and manage information, but only in a purely arbitrary way, revealing more the unique self than universal truths.

Montaigne remained content with this revelation, which finds remarkable currency in our own age of self-awareness (or, as some might say, self-absorption). But few contemporaries shared the serenity of his outlook, for most lacked a sense of self strong enough to serve as an existential anchor in a sea of diversity. Accordingly, his new organization of knowledge was destined to remain highly idiosyncratic, providing contemporaries with little to offset the skepticism he had unleashed. Notwithstanding his geniality, levelheadedness, and open delight in all things human, Montaigne cast a dark shadow across the seventeenth century that his successors struggled to dispel.

Instead of cherishing the irreducible complexity of the world, the early modern mind took greater comfort in a more radical solution to the breakdown of commonplace thought, one heralded by René Descartes (1596–1650). Far from marveling at diversity, he preferred to discard it and start over, basing knowledge on firmer foundations than had his humanist predecessors. In the *Discourse on the Method,* he describes these foundations as "clear and distinct ideas." This famous expression has technical meaning in Descartes's philosophy, but in general it signifies knowledge characterized by the same degree of certainty found in mathematics. Although his philosophy was not specifically mathematical, Descartes would lay the foundation for an analytical vision of knowledge based upon mathematics, in which symbolic language would supplant natural language as the most effective means for organizing and managing information.

With his radical solution to the breakdown of commonplace thought, Descartes sought to bridge the gulf between what we would today term the sciences and the humanities, thereby creating a more useful and inclusive organization of knowledge. This gulf had not existed for the scholastics because they regarded *scientia* (exact knowledge) as leading to *sapientia* (divine wisdom), which naturally encompassed all forms of human wisdom. The unity of thought first began to come apart in the Renaissance, when the humanists championed the moral benefits of rhetoric over the intellectual ones of logic.

Skepticism about the utility of his education led Descartes to question this humanistic assertion, after which he attempted to formulate a new kind of *scientia* that would lead to a new kind of *sapientia,* an all-embracing human wisdom. Excluding from the latter realm all subjects open to the slightest doubt, he sought to create a core of knowledge more useful than that of the humanists, with which to make a brave new world rather than to revive an ancient one. The breadth and depth of his bold analytical vision of reality will be the theme of the next chapter, but we shall close this one by revealing how his desire to reorder the storehouse of knowledge originated in response to the breakdown of commonplace thought.

Descartes attended the Jesuit school at La Flèche, where he received an essentially humanistic education, with heavy emphasis on rhetoric. The Jesuits utilized the humanist curriculum to create eloquent spokesmen for the church, training their students in classical language and literature by means of commonplace books. They augmented this rhetorical education with upper classes in logic, mathematics, and metaphysics. But these studies, too, were cast in a rhetorical mold, designed to train students to defend accepted authorities, chiefly Aristotle. Although the rhetorical thrust of humanistic education had remained unchanged for generations, attitudes toward the world had become progressively more complex, characterized by an acute awareness of historical and cultural relativism, an awareness exacerbated by printing.

The ease with which the young Descartes dismissed his education indicates the extent to which relativism had undermined commonplace thought by the early seventeenth century. In the autobiographical portion of the *Discourse on the Method,* he describes his youthful dissatisfaction with a humanistic education that had failed to fulfill its promise of moral guidance. Far from furnishing the mind with norms, classical literature, especially history, served only to highlight for Descartes the differences between antiquity and modernity. Too much immersion in classical culture, he observed, made one a stranger in one's own world.

Even the philosophical curriculum, which augmented the humanist one in Jesuit education, failed to offer the young Descartes any useful, worldly orientation. Like Montaigne, he complained that the opinions of philosophers were too diverse and contradictory to provide a stable foundation for knowledge. Only mathematics offered him any sort of clarity at all, and thus he delighted in it. But initially he did not conceive of mathematics as having any application beyond the mechanical arts, none of which were conceptually robust enough to provide him with a point of orientation amid diversity.

The young Descartes yearned for certainty, not philosophical or scientific but moral certainty, affording the kind of worldly wisdom his humanistic education had promised. Having failed to find it in books, he sought it in travel, serving in both the Protestant and Catholic armies then maneuvering against each other in Holland and Germany. Travel, though, only confirmed his sense of cultural relativism, without providing him with any moral certainties. At this point of confusion in his life, he was once again drawn to the clarity of mathematics, which he began to pursue in a number of diverse studies: on falling bodies, on the pressure of liquids, and on music as a science of proportions. In these studies, he was fortunate to find a mentor in Isaac Beeckman, a Dutch savant with a special interest in mathematics and natural philosophy.

From his mathematical studies during 1618–19, Descartes concluded that *scientia* would have to show the way to *sapientia,* that true wisdom was a function of the proper ordering of knowledge. He thus began experimenting with various schemes of order to supplant that of his education. Raymond Lull, the medieval author of a universal art for finding the truth, offered one such scheme, and Descartes became intrigued enough by Lull's mystical art to ask Beeckman whether it was worth pursuing. In a similar vein, Descartes may have also flirted briefly with the Rosicrucians, who claimed possession of the mystical key to knowledge.

More revealing is his interest in Lambert Schenkel's *Art of Memory.* This late-sixteenth-century treatise largely rehashed traditional mnemonic schemes, which disposed of knowledge in a network of places arranged like the rooms of a building or the buildings of a city (see plates 9 and 10). Each place contained images designed to trigger the memory, with the component parts of each image keyed to the different kinds of information stored in the place. A place might contain, say, the personified image of "Grammar," a woman whose figure, dress, and stance served as mnemonic hooks for specific kinds of information about the liberal art she embodied (see plate 11). Additional information might be inscribed on the image by means of "visual alphabets," easily remembered objects whose shape or pronunciation recall letters, thus compressing alphabetically stored information by visual means (see plate 12).

In the manner of commonplace thought, Schenkel's mnemonic structure represents not so much a hierarchy of knowledge as an assemblage of discrete pieces of information, each of which derived its significance from the moral norms everyone took for granted. Perhaps sensing the need for a more substantial framework to hold this structure together, Schenkel made veiled reference to an occult mnemonic scheme for divining the mystical unity of knowledge, which he attributed to celestial influences. In the sixteenth century, the occult offered a popular means of intellectual closure, revealing what Schenkel termed the "causes" that underlay and explained an increasingly complex world.

The comments Descartes made in a notebook after reading Schenkel reveal his yearning for order amid the breakdown of commonplace thought. He too desired closure, though the "causes" he sought were evidently not of Schenkel's mystical sort: "When one understands the causes, all vanished images can easily be found again in the brain through the impression of the cause. This is the true art of memory, radically different from [Schenkel's] nebulous art." Descartes's mnemonic system would offer closure by showing how all causes ultimately reduce to one. He envisioned an intellectual system in which knowledge would be based on a single, firm foundation, from which the mind could extend outward, "forming images dependent upon one another," thus creating a unified body of knowledge. His comments, though, suggest the classificatory nature of this closure, in which species reduce to their genus as the mind ascends the hierarchy of the sciences.

Descartes noted in passing that Schenkel's art "uses up too much paper," a comment underscoring his interest in mathematics, which would eventually show the way toward a new science of order. We shall examine this science next and see how it came to be expressed in a symbolic language of analytical abstraction. This language lies at the heart of what we shall call the "analytical vision of knowledge." The relations expressed in this new language would eventually replace the classificatory abstractions of natural language as a means of organizing knowledge and managing information.

As we explore the analytical vision in the next chapter, let us keep in mind Descartes's yearning for order amid the breakdown of commonplace thought. In his attempt to establish a new order embracing the human world as well as the natural one, he would unwittingly divorce exact knowledge (*scientia*) from wisdom (*sapientia*), fostering two distinct intellectual cultures—the scientific and the humanistic—that would remain irreconcilable until our contemporary information age.

CHAPTER FIVE

Numeracy, Analysis, and the Reintegration of Knowledge

There is no problem that cannot be solved.

—FRANÇOIS VIÈTE, *Introduction to the Analytic Art*

Descartes's Dream

We all dream, sometimes when awake, other times in our sleep. Yet while our dreams frequently have importance to us personally (or to our analysts), they seldom portend much for the fate of humankind. Unlike our ordinary reveries, a singular dream of René Descartes's augured such significance, even suggesting to a later commentator that "the modern world, our world of triumphant rationality, began on November 10, 1619, with a revelation and a nightmare."

With great detail Descartes recorded his dream in three parts. In the first two parts, he noted being pushed about by a violent whirlwind, struggling to reach sanctuary in the College chapel at La Flèche, being offered a melon, and painfully awakening, only to drift back into a horrific vision of thunder, lightening, and sparks flying about the room. Sweats and chills. By contrast, in the third part he recounted a calmer image, in which he saw on his table a dictionary and a poetry anthology, the latter opened at a passage of the Roman poet, Ausonius (A.D. 4th century), which asked: "What path shall I follow in life?" A stranger then presented him with a bit of verse containing the words "yes and no," and shortly thereafter, still half asleep, Descartes began to interpret his own dream, an interpretation he immediately recorded in his diary upon awakening. The dream, he believed, had warned him away from the errors of his past life and had given him his vocation, which was nothing less than the discovery of the "foundations of the marvelous science."

Although not in complete agreement, scholars have usually seen this epiphany as a grandiose vision of the unity of mathematical science, a vision surely affecting the fate of humankind. Ample evidence suggests Descartes saw it that way too. When his dream occurred, the twenty-three-year-old gentleman-soldier-student had already been working for over a year with Isaac Beeckman, whom he considered the "inspiration and spiritual father" of his studies. In the next decade he pursued the investigations that would culminate in the creation of coordinate geometry (what we now call analytical geometry), a new discipline unifying two mathematical subjects, geometry and algebra, each of which previously had been considered a separate branch of knowledge. The same decade also witnessed the beginning of his philosophical reflections on these investigations and of his concerns for a method "of rightly conducting the reason and seeking for truth in the sciences."

Descartes first articulated his reflections on method in the *Rules for the Direction of the Mind,* composed in the 1620s, though not published until 1701, long after his death. In that work he sharply criticized the traditional, classificatory pattern of thinking embraced by his teachers and predecessors; he proposed in its place a new kind of intellectual abstraction, one based on mathematics. He named this new thinking *mathesis universalis,* "universal mathematics," and claimed it would yield the "order and disposition of the objects toward which our mental vision must be directed if we would find out any truth." Properly understood, universal mathematics would provide, in today's parlance, the global foundations for anything we could know.

Another feature of his potent dream announced a second and equally important dimension of Descartes's vocation. We gather from his early biographer Baillet that the image of the poetry anthology represented "philosophy and wisdom linked together," the vision of the marvelous science. Yet in his dream Descartes also saw a dictionary on his table. What did it signify? Baillet suggests "all the various sciences grouped together." Not only would there be a new foundation for knowledge, there would be a new compilation and ordering of it as well. With this single image Descartes bridged the worlds of the medieval *summa* and the modern encyclopedia.

We should remember that besides meaning "exact" knowledge, the Latin *scientia* had also referred to "ordered" knowledge of natural things in its pairing with *sapientia,* the ordered knowledge of divine things. Throughout the seventeenth and eighteenth centuries science would gradually acquire its more modern, restricted sense and refer increasingly to the knowledge derived from a systematic method of discovery based on mathematics and experiment. Descartes contributed mightily to the evolution of scientific method. At the same time he never abandoned the broader meaning of science as organized knowledge.

To the end of his life he believed his mathematical vision would provide a new order, eventually embracing not only the natural sciences but also the moral ones. The old order had been swamped by an overwhelming array of historical forces: the emerging Copernican revolution in science, the discoveries of new worlds and peoples, the fragmenting of Christian truth, the skeptical relativism afflicting humanistic studies, and the information explosion born of printing. No one captured the intellectual panic at drowning in a sea of unorganized, unreliable, and hitherto unknown information better than the poet John Donne, writing in 1611:

> And new philosophy calls all in doubt,
> The element of fire is quite put out;
> The sun is lost, and the earth, and no man's wit
> Can well direct him where to look for it.
> And freely men confess that this world's spent,
> When in the planets and the firmament
> They seek so many new; they see that this
> Is crumbled out again to his atomies.
> 'Tis all in pieces, all coherence gone;
> All just supply, and all relation.

Brash in his optimism, Descartes thought that two or three generations would yield ample time to set aright this state of affairs, to find coherence once again, to accumulate and reorganize all the knowledge humankind needed, and to reunify knowledge with wisdom.

What is there about mathematics—so forbidding and mysterious for so many—that animated Descartes's unbounded confidence in it? Why did he and others after him believe that the "imagination mathematical" (as the sixteenth-century mathematician John Dee had christened it) was capable of coping with the deluge of information so troubling to the order-seeking minds of the age? Natural language and the classifying impulses it had shaped over the centuries proved themselves incapable of responding to the crisis. Overflowing with information, the old classificatory categories had become useless for discernment or discovery amid natural and social flux. Why would mathematics fare better as ground and organizer of human knowledge?

The brief answer, which we shall explore and develop at some length, is that the emerging, "modern" mathematics provided a new symbolic language whose principal advantages in mastering information lay in its greater abstraction from the universe of words and things, from experience, and in its unprecedented capabilities of analysis. Of course, all written language is symbolic, at least in the sense that letters and words stand for sounds and objects. But modern mathematics is *purely* symbolic, with its own abstract symbols for components and operations, and with its own logic, procedural rules, and connections to the world. Its products draw us farther away from experience than does literacy, letting us see and shape the world from a totally different, critical perspective than does classification.

All mathematics is based on the cardinal and ordinal principles of number, themselves tracing back to the age-old features of correspondence and recurrence manifest in early counting practices. Once captured in purely symbolic form, these two principles became the foundation of the new information technology that we identify as modern numeracy. And numeracy, in turn, generated a new information idiom, a new way of giving form to human experience: the analytical vision of knowledge.

In the classical age alphabetic literacy had arisen as an information technology enabling the Greeks to formulate the idiom of wisdom, itself an outgrowth of the classifying capabilities of natural language. Now in the modern age, the information technology of numeracy would likewise elicit the information potential of pure, abstract number. The new idiom of knowledge was understood to be the product of analysis—breaking things down into their constituent parts, capturing the parts in abstract symbols, and connecting the abstract symbols in forms or formulas. This idiom would enable humankind to identify new sorts of information, to discern true from false information, and to manage information in ever-increasing quantities.

Descartes did not apply the term information technology to his universal mathematics; this is our reading, about which we shall say more later on. Nevertheless, his and others' growing confidence in modern mathematics stemmed precisely from the fact that, as a new information technology and idiom, mathematics gave the age an answer to the information overload of its day. With it the book of nature would be rewritten, a multi-authored tome by men (mostly) of genius and vision, who both created and spoke the new language. Without it, Galileo warned, one would forever wander "about in a dark labyrinth."

Many seventeenth- and eighteenth-century savants came to believe that the new technology and the analytical vision it sired would eventually reintegrate new learning with old and satisfy their passionate longing for a unified order of knowledge and wisdom. Among them, Descartes glimpsed as much in his dream, then spent his life trying to make it happen. Like so much of our history, this process was halting and laden with ironies. Chief among them, the analytical vision itself would eventually undermine the very substantive, moral order of *sapientia,* wisdom, that Descartes had intended to restructure with the tools of universal mathematics. Within this ironical tale, as wisdom quietly gave way to knowledge, the shape of information was transformed from the classifying of words and things to a mathematical mapping and ordering of the world. The modern information age was born.

The Hallmarks of Early Numeracy

As denizens of modern, numerate culture (wherein even arithmephobes see everywhere the numeracy they fear), it is difficult to step outside our times in order to appreciate the far-reaching scope and depth of the intellectual ground shift that produced the modern age. To do so we must first digress briefly and introduce three central themes of premodern numeracy whose significance bears directly on the emergence of the analytical vision: (1) the nature of mathematical abstractions themselves; (2) the sharp dichotomy between arithmetic and geometry; and (3) the rise in Western Europe of a fully positional, symbolized number system, with symbols standing for operations as well as things. These themes help us appreciate how the components of modern numeracy first coalesced into a full-blown information technology during Descartes's lifetime.

The earliest and most primitive evidence of numeracy lies in the use of tokens, those primordial signs that predated writing and that gave humankind its first means of record keeping. From its beginnings in token iteration, numeracy evolved gradually into a sophisticated and systematic ability to enumerate or count, or more generally, to engage in quantitative thinking. In so doing, it revealed the two fundamental features that pervade all mathematical abstractions and that would anchor the modern age's information technology: correspondence and recurrence. These characteristics ultimately became known as the cardinal and ordinal principles of abstract number, principles derived from the late-nineteenth- and early-twentieth-century investigations into the logical foundations of mathematics. But long before they acquired the elevated status of principles, the cardinal and ordinal features of number were utilized unreflectively in early counting systems. In fact they provide the signposts by which we recognize a numerate culture.

The basic correspondence between a token (or more generally, a "unit-symbol") and the object it stands for expresses the central idea of the cardinal concept of number. Taken strictly by itself, it does not imply counting, only making one-to-one match-ups between the discrete members of different collections. A theater, for instance, contains some seats; people enter and claim them, one seat per person. If there are neither standing people nor empty seats, then the cardinal numbers denoting the collections of people and seats are equal. If there are empty seats, the seat collection is larger than the people collection, and vice versa if standing people. Note, we have not counted or organized the members of these collections serially in any way, only correlated seats and persons. Thus cardinal number refers simply to a total collection of objects, correlated with another collection and identified with a number word.

Although little understood until well into the modern period, the cardinal correlations of number differ fundamentally from the referents of language and classifying, even while they both probably stem from the same practice of making visual comparisons. Through the medium of sight, the mind derives from comparisons of several collections of things a numerical, mental extract—'seven', for example—which corresponds to the same number of objects in a collection (whether sheep, sins, or wonders of the world is irrelevant). In a similar visual process, the mind elicits a classifying term, say 'vertebrate', whose defining properties are then understood to correspond to any animal with a backbone.

Both numerical and classifying terms are created by an act of visual comparison. But the results of the act differ. The term 'vertebrate' itself is defined by its visual content; the content prods us to imagine a backbone, and armed with this image we correlate an observed object and a category. The term 'seven', however, carries no such visual content. In imagining seven we really have no image at all. We may imagine seven fingers or pebbles or sheep, but these are all groups of specific objects, not 'seven' per se. The objects serve as visual placeholders or reminders for completely abstract, nonvisual information units, which can be correlated with any things whatsoever.

Cardinal correlations, though necessary, cannot suffice for a counting system. To count, we need not only a means of correlating unit-symbols with objects but also the capability of ordering them in some kind of sequence or series. Ordinal numbers do this. Whereas a cardinal number refers to the total number of members belonging to a collection, the ordinal number designates the ordering of those members within the collection. Ordinal number, in short, refers to a number's rank. When '250' indicates the total number of seats in a theater, it is a cardinal number, referring to a collection. When '250' designates a reserved seat in the theater, it is an ordinal number, identifying one specific seat because of its serial ranking in the collection of seats.

As counting evolved, ranking was achieved by devising an ordered series of numbers that "progresses in the sense of growing magnitude, the natural sequence: one, two, three . . ." The natural sequence means that one can always follow a number with its unit successor and that one can repeat the process ad infinitum. This is the principle of recurrence. Reaching and managing larger collections of things were made possible by devising a "collective unit" to serve as a base of the counting system. Repetition of the base enabled the rank ordering and calculating of indefinitely large numbers. Historically, all counting systems have coalesced around bases 2, 5, 10, 20, or 60, the most widespread being our familiar denary or decimal, base-10 system.

Throughout the ancient world, counting grew to be intricate and sophisticated, with many and varied techniques devised for the basic operations of addition, subtraction, multiplication, and division. In some cultures manipulating fractions, computing rapidly with an abacus or counting board, and figuring powers and roots (squares, cubes, and the like) became commonplace. The number realm expanded from whole numbers to fractions to rational numbers, and even to the irrational numbers. Underlying all these developments are the cardinal and ordinal features of number, and their implicit principles of correspondence and recurrence.

These same principles pervaded another facet of early quantitative thinking, geometry, at which the Greeks excelled. Yet, curiously enough, the Greeks came to consider geometry and arithmetic as completely separate realms of knowledge, as different and irreconcilable as Athens and Sparta. Overcoming this separation would be the key to the Cartesian achievement of transforming early numeracy into universal mathematics.

We can readily identify the cardinal and ordinal features of number in Greek geometry. Calculations, for instance, often required a one-to-one correspondence between numbers and line segments, such as obtaining the hypotenuse of a right triangle with sides of 3 and 4 units of length. And numbers themselves were often considered as geometrically arrayed points, as with linear, square, triangular, and oblong numbers (see figure 5.1). At the same time, the ordinal feature of recurrence was manifest in the treatment of continuous lines and of the infinity embedded in them. For Pythagoras, the diagonal of a square exemplified a finite line segment, but one whose length cannot be expressed as a rational fraction, a finite ratio of two integers. Rather, as he showed from the famous theorem that now bears his name, the numerical ratio producing the di-

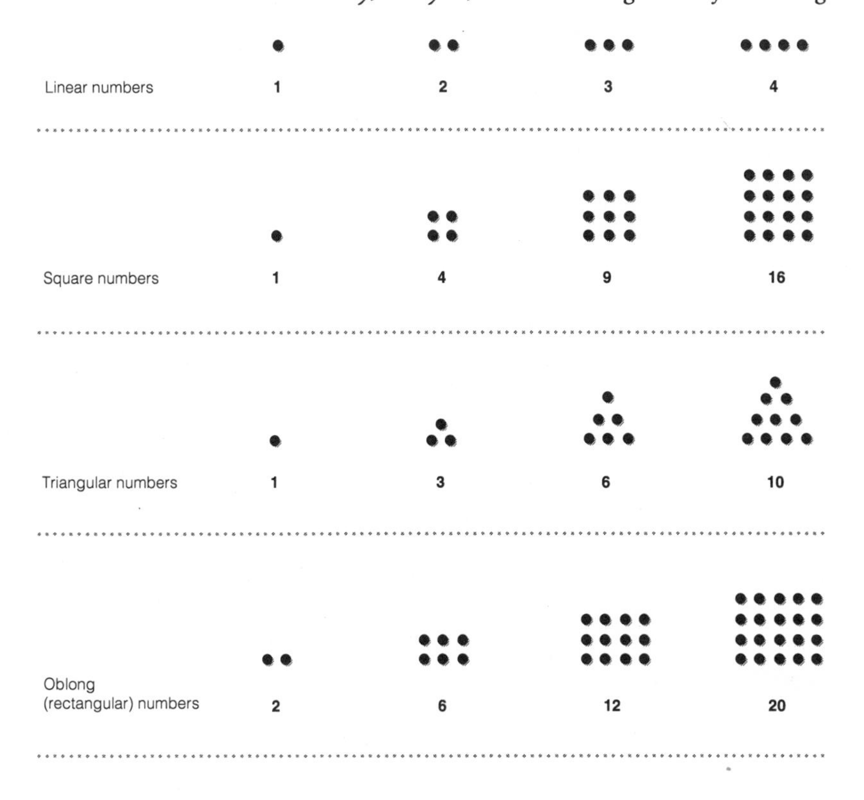

FIGURE 5.1. Pythagorean Figurate Numbers

agonal led to unending repetitions, infinite divisions, which we now designate as $\sqrt{2}$, a notation the Greeks did not have. Thus were discovered the "irrational" numbers, and with them the mind-boggling idea of infinity, so horrific, legend has it, that Pythagoras and his followers kept it secret for fifty years.

The stumbling block of infinity prevented the union of geometry and arithmetic by posing a problem. How could finite, continuous lines be reconciled or correlated exactly with an infinity of discrete numbers or ratios or points? The Greeks had no answer. Nor did anyone else in the ancient world. From this problem evolved the notion that separate perceptions of continuity and discreteness produced distinct and incommensurable categories or genera or fields of knowledge.

Put another way, mathematical abstractions themselves were subordinated to the classifying procedures of language and literacy. One began with an image of continuity, a perception common to things like lines, spaces, shapes, objects, figures, time, and place. For Aristotle continuity possessed, further, the identifying mark of a "common boundary" that joined a continuous whole (say a line) and its parts (the segments that comprised it). This feature made all continuities "infinitely divisible," since lines forever divided in half would still be lines — short ones, but lines nonetheless. Phenomena identified as having common boundaries among their parts and as being infinitely divisible were then subsumed under 'geometry' as the category of mathematical knowledge that treated continuous spaces.

In contrast, discreteness was understood as a commonly perceived property of distinct numbers, individual objects, collections of units, counting, and the like. This perception permitted placement of all such objects in the class of arithmetical phenomena. Aristotle actually claimed that both number and spoken language were subclasses of 'discreteness' because neither numbers nor word syllables could be joined "together at a common boundary." Each number and each syllable remained self-contained, separate, determinable. Four anythings were always demarcated as 'four', never 'five', whereas line segments and points could never be so demarcated.

From this account it followed that numbers could not stand in a direct, one-to-one correspondence with the points on a line. Accordingly, continuous lines or spaces remained incommensurable with discrete, self-contained numbers, and so too the general categories of arithmetic and geometry. Once again the master classifier, Aristotle: "Arithmetic is about units and geometry is about points and lines," and one cannot "prove anything by crossing from another genus — e.g., something geometrical by arithmetic."

Despite the fact that the cardinal and ordinal features of number permeated both arithmetic and geometry, therefore, the Greek classifying mind kept these disciplines separated from one another, with the momentous consequence that mathematics would remain for centuries pigeon-holed in taxonomical thought. Distinct genera of the yet more general category of quantity, arithmetic and geometry were later codified in the curriculum of the Middle Ages as parts of the quadrivium, with arithmetic and geometry comprising the theoretical branches of mathematics, music and astronomy their practical applications.

With the hindsight of over two millennia it seems apparent now that the inherent restrictions of early numeracy are linked significantly to the absence of a fully developed, positional system of symbol notation and operations. Such an absence impeded mathematical resolution of the problems surrounding infinity (with some notable exceptions). It also limited geometry to methods of synthetic construction and proof (notwithstanding the depths of Greek insights). Overcoming these impediments became possible only with the arrival in Europe of a positional counting system and the standardization of operational symbols and notation. These developments framed the beginnings of a purely symbolized language that would soon be relied on as the most assured means of knowing the natural world.

Considered by some "the most successful intellectual innovation ever made on our planet," the remarkable invention of place-holding or positional counting comes to us from India, where it had evolved out of early numeracy dating from the third millennium B.C. The Indian system combined several advantages only suggested by other early counting systems. It was denary or decimal throughout and possessed a limited number of unique and abstract symbols for the digits 1 through 9. It utilized place-value notation, wherein the value of a digit was completely determined by its position in an array of digits. As in modern decimal counting, a '7', for instance, could mean 7 units, 70 units (7 × 10), 700 units (7 × 10 × 10) and so forth, with the same numeral deriving its quantitative value from its place in the ones, tens, or hundreds column. (This contrasts with other, early counting systems, which generally were "additive," meaning that the value of any number was determined by totaling its numerals, regardless of their position.)

Further, the Indian system incorporated the concept of zero in number word and symbol. The cipher for zero stood as a placeholder in empty columns and, more important, it carried the idea of nothing (originally the "void") in operations, including subtraction. Place-holding and the zero permitted borrowing and carrying remainders from column to column in the now familiar way, greatly facilitating calculations. The number notation that eventually developed into our modern number symbols also originated in early Brahmin texts.

Throughout every major cultural confrontation in which it figured, Indian positional counting and notation asserted itself, surpassing rivals. To the west, during the time of early Islam, the Indian system gradually prevailed upon an Arab culture that already possessed a sophisticated numeracy of its own. The number symbols for integers acquired the forms we still use today (with some later changes to the characters for five and zero), and Indian calculating techniques were adopted as well. Through frontiers of cultural and commercial contact (Spain, Sicily, Southern Italy, Constantinople), the system filtered into Europe during the Western Middle Ages, when the Moslem world exercised political and cultural hegemony throughout the Mediterranean and the Levant, from Toledo to Baghdad.

By the fourteenth century the Indo-Arabic system was being widely used in European commerce and studied in schools. From then through the seventeenth century its presence grew steadily, although not without some opposition from merchants who feared fraud with the new, easily altered numbers. Die-hard "abacists" continued to record their figurings in Roman numerals as well. The abacists computed (quite rapidly) by moving counters on an abacus or counting board, and for a while outpaced the initially awkward and slower "algorithmists" (see plate 13). The latter calculated with the new number symbols on slate or blackboard or eventually on paper, just as the Indians had written numbers in sand with sticks or had traced out symbols with sand, flour, or seeds.

Nevertheless, however rapid the computations, the abacus could not store as much numerical information as could chalk and slate, or pen and paper. This brute fact joined with other advantages to help the algorithmic techniques attain widespread popularity: the increasing availability of paper, the ease of correcting and erasing mistakes, the expanding demand for more sophisticated calculations, the growing commercial familiarity of the new system, and ultimately the sheer facility of calculating (especially accounts in double-entry bookkeeping, with their straight columns and rows). Even so, the new methods advanced but gradually; not until Shakespeare's day did Roman numerals finally fade from commercial use in Europe. By then the merchant of Venice was weighing his pound of flesh in Indo-Arabic notation.

In keeping his accounts, the merchant was also likely to be using several new, operational symbols of a sort previously unknown. More conjecture than evidence persists as to the precise origin of many arithmetic symbols, including the modern plus (+) and minus (−) signs. Whatever their origins, whose intriguing details would take us too far afield, by the end of the sixteenth century these two symbols had become conventional notation for the arithmetic operations of adding and subtracting. Other, equally important operational symbols became standardized during the following century, including the '×' for multiplication, the '÷' of division, the equals sign (=), and the symbols for greater and lesser (<, >). Finally, the modern symbol for root (√), invented in Germany, also gained popularity gradually throughout the seventeenth century. Prior to that time roots were more often designated by an R (from the Latin *radix*, "root") or sometimes an ℞ (from which our modern symbol for prescriptions; apothecaries formerly mixed "root" elements to make compound drugs).

The sixteenth and seventeenth centuries witnessed the burgeoning of modern mathematical symbolism. As a random sampling, we have only to peruse pages of the first English-language arithmetic from early in the period (*The ground of artes* [1543] by Robert Recorde) to see a lengthy and lumbering (to the modern taste) rhetorical account of arithmetical operations, even while it uses Indo-Arabic numerals. Compare it with a later work such as that of Brook Taylor, whose *Methodus incrementorum directa et inversa* (Method of Direct and Inverse Increments, 1715), although still in Latin, fills page after page with formulas, series, operations, and demonstrations, all expressed in recognizably modern, symbolic notation. Taylor's rhetoric serves only to introduce the notational mathematics, not to articulate the problems.

These works provide striking testimony to the nearly two centuries of symbol development and standardization that elapsed between them, an era of unprecedented transformation in humankind's ability to think logically in abstract symbols. Scholars generally agree that during this period in Western Europe there emerged what we usually call modern, abstract mathematics (often termed "relation-mathematics"), as opposed to premodern, concrete mathematics (or "thing-mathematics"). Early numeracy, keyed to collections of things, was transformed into a numeracy centered on operations with abstract symbols and based on abstract number. In acquiring its own language and rules, modern, abstract mathematics blossomed into a fully developed information technology, one that made the accompanying advances in scientific knowledge "revolutionary." Fixing the conventions of symbolic notation provided a major step toward this accomplishment.

Viète and the Beginnings of Modern Analysis

While arithmetic calculations were being simplified through a positional number system and new operational symbols, arithmetic itself was metamorphosing into a language of far greater abstraction, that of symbolic algebra, which would play a crucial role in the formation of modern numeracy. Algebra per se is not an innovation of modern mathematics but refers to a generalized method of treating arithmetic calculations, particularly those involving equations with unknown quantities. The term itself comes to us from the title of a treatise by the Arab mathematician Mohammed Al-Khwarizmi (c. 780–859), *Hisab al-jabr w'al-muqabala*, which was translated into the Latin *Liber algebrae et almucabala* by Robert of Chester in the twelfth century, whence the English word 'algebra'. (Latinized, Al-Khwarizmi's name also gave us 'algorithm'.)

The Arabic title of Al-Khwarizmi's treatise translates literally as "the calculation of reduction and confrontation," with *jabr* and *muqabala* denoting, respectively, the practices of transferring negative terms to the other side of an equation (thus in modern notation, $a = x - 6$ becomes $a + 6 = x$) and canceling similar terms on both sides of an equation (thus $a + b - 6 = x - 6$ becomes

$a + b = x$). *Jabr* also designates the practice of setting a broken bone, the origin of today's medical use of the term 'reduction'. Employing such techniques reduced complicated problems in order to isolate unknown quantities and to discover their known equivalents.

We find many variants of general "algebraic" methods in early counting systems, and many of these involve roots, powers, and other multiples (or coefficients) of unknowns. For the most part, these earlier problems were written out and expressed rhetorically. Even Al-Khwarizmi wrote his "equations" in words. (The difficulty of imagining an equation without seeing an equals sign offers another indication of just how thoroughly abstract symbols imbue our modern numeracy.) Most historians assign the term "rhetorical algebra" to premodern algebraic methods. After Al-Khwarizmi's works reached Europe, many Renaissance and early modern mathematicians developed further techniques for expressing and manipulating various classes of equations. In the process they inched from rhetorical to "syncopated algebra," where basic problems are still written out, but where some notational shorthand occasionally abbreviates rhetorical repetitions.

The creation of a fully symbolized algebra comes to us largely from the French mathematician François Viète (1540–1603), who first suggested and devised rules for a "general letter algebra" in his *Introduction to the Analytical Art* (1591) and subsequent works. A highly placed lawyer in the court of King Henry ("Paris is worth a mass") the Fourth, Viète displayed a keen mathematical imagination, one that ranged from deciphering encoded enemy (Spanish) dispatches to solving equations of higher degrees and various problems of trigonometry.

But his real passion and most cherished intellectual occupation lay with the "threefold analytic art," the *logistica speciosa* ("species calculation") whose aim was to improve on the arithmetic *logistica numerosa* ("numerical calculation") of the Greek mathematician Diophantus (A.D. 3d century). Viète's inspiration derived from Euclid's theory of proportions, which he saw as the cornerstone of mathematical structure. Yet unlike Euclid, whose techniques were synthetic—relying on axioms and theorems to construct geometrical figures and proofs—Viète introduced analytical techniques for calculating magnitudes.

He began by revising a principle called the "law of homogeneity," the "first and supreme law of equations and proportions." A Greek precept of ancient standing, the law stated that only the same kinds of quantities could be compared with one another. Aristotle had explained that the two main species of "quantity"—arithmetic ("numerable" quantity or "plurality") and geometry ("measurable" quantity or "magnitude")—differed as apples to oranges and were therefore incommensurable, incomparable. Viète showed that Aristotle was mistaken, that geometrical magnitudes could be structurally manipulated as powers of arithmetic quantities, and that numerical and geometric proportions could be compared.

Viète accomplished this by developing the notion of "scalar magnitudes"—a new idea at the time, but one that, because of its expression in traditional, generic terms, still jars our modern mathematical sensibility. These magnitudes, he defined, "by their own nature ascend or descend proportionally from genus to genus." We begin with a "side or root." A side (x) multiplied by itself makes a square (x^2, or *quadratum*, from which 'quadratic'); a square times a side produces a cube, both in the geometric and (we can now say) algebraic sense, x^3; a cube times another side yields a "squared square" (*quadrati-quadratum*), or fourth power, x^4. More generally, a length times a breadth produces a plane figure, a plane figure times a side produces a solid figure, and so on.

With scalar magnitudes Viète created a hierarchy of proportions, effectively uniting the previously incomparable genera of geometry and arithmetic: "(1) Length and breadth; (2) Plane; (3) Solid; (4) Plane-plane; (5) Plane-solid; . . . (9) Solid-solid-solid"; and so forth. Viète termed each of these subcategories a "genus," but the relations between them were proportional, not classificatory—relations of degree rather than kind. (In today's expression, his arrangement is an ordinal sequence of powers: x^1, x^2, x^3, x^4, . . . , x^n.) On this basis he then treated the proportions of spatial magnitudes as functions of arithmetic powers, the practice he called numerical calculation, *logistica numerosa.* In short, Viète assumed the proportions of both geometric magnitudes and arithmetic powers corresponded to one another in a one-to-one fashion.

The capstone to Viète's endeavors was to generalize from numerical calculations to the whole category of quantity, the *logistica speciosa.* This meant reckoning "by means of species or forms of things, as, for instance, the letters of the alphabet." How can one generalize from numerical calculations of spatial magnitudes? Let letters of the alphabet serve as abstract characters in place of numbers. This had nothing in common with the earlier Greek (and Hebrew) convention of using the alphabetic sequence of letters to stand for specific numerals in their natural series, one, two, three, . . . Viète assigned his letters arbitrarily to correspond in a one-to-one manner with *any* number whatsoever. Thus he introduced problems with such phrases as "Let the given number be D, and the difference be B."

Once letters could correspond to any number in general, the arithmetic operations between letters, articulated increasingly in symbols (+, −, ×, ÷,

√, =, etc.), permitted manipulation of entire classes of numerical problems in purely symbolic terms. This technique far surpassed the use of symbols as shorthand abbreviations in the accounts of earlier rhetorical and syncopated algebra. As containers of numerical information, the symbols themselves were now singled out and made the object of further reflection—revealing yet again the mind's natural tendency to rework the products of its own creation. From this reflective turn, modern numeracy would begin its flight into heightened abstractions. And from Viète forward, analysis would come to mean, in large part, the reduction of complicated mathematical relations and calculations to the equations of their general algebraic form.

As remarkable as they appear from our perspective, Viète's innovations must be viewed in their historical context. His continued use of the categories of genus and species suggests a classifying mind trying to incorporate the new insights of symbolic and analytical techniques into a deeply imbedded intellectual architecture. True, he did assume a one-to-one correspondence between the proportions of arithmetic powers and spatial magnitudes and in so doing challenged the centuries-old division between arithmetic and geometry. But the powers of spatial magnitudes beyond solids (or cubes) really had very little geometrical meaning in a universe still composed of Euclid's three dimensions. (Viète was not here envisioning later space-time geometries of four or greater dimensions.) The novel and incisive move of positing a one-to-one correspondence between letters and numbers in his algebra did not extend effectively to similar correlations between numbers and points.

Despite his insistence on the technical commensurability of proportions in arithmetic and geometry, therefore, this part of Viète's project was stillborn. A different sort of program was needed to devise the techniques for mapping numbers onto lines, for making an algebraicized arithmetic thoroughly commensurable with geometry, and for creating a fully abstract information idiom, capable of connecting with the world of experience. It awaited Descartes.

Descartes's Analytical Achievement

Inspired by his dream of a universal mathematics, Descartes launched his own mathematical studies from "where he [Viète] left off." As his project got underway, he began inventing the techniques that enabled him to transform the emerging, modern numeracy into a fully symbolized language far exceeding anything his mathematical forebears had imagined. The new techniques made it possible to articulate, store, manipulate, and transmit exact information about phenomena by means of completely abstract symbols. In this abstract language of universal mathematics lay the heart and soul of the analytical vision. It gave Descartes a new and powerful means of recasting *scientia,* cognitive order, and, he remained convinced, of reintegrating this order with *sapientia,* wisdom.

Beginning with his earliest description in the *Rules for the Direction of the Mind,* Descartes presented universal mathematics as a single discipline, a "general science" devoted to "discovery of an order." Order was to be achieved by separating simple from complex "facts" in any series and by marking the "interval, greater, less, or equal," between them. Herein lay the "chief secret of the method," which, if properly utilized, would yield the "correlative connection and natural order" of cognitions.

He spoke adamantly, even haughtily about this new order. The entire science of "pure mathematics" embraced the "form" of "all questions" concerning "proportions or relations of things." Mathematics itself would use the new language of symbols and display the terms of a problem abstractly, in a "detached and unencumbered way." To prevent wasting intellectual powers, problems would be put down on paper, rather than be committed to memory. On paper one could state how "the abstract formulation is to be made and the symbols to be employed, in order that, when the solution has been obtained in terms of these symbols, we may easily apply it, without calling in the aid of memory at all, to the particular case we are considering."

With these and similar expressions Descartes forecast both the new analytical order of knowledge and its connection to the world. His words enunciated the cardinal and ordinal features of number, those hallmarks of numeracy and modern analysis. The new learning, he stressed further, bore no affinity whatsoever with any "ontological genus such as the categories employed by Philosophers in their classifications." In effect, he announced that our experience of the world would no longer be informed by means of natural language, but through the symbolic language of pure mathematics.

Descartes's breakthrough to new analytical possibilities lay in uniting Euclidian geometry and symbolic algebra into a single system of coordinate geometry. About the same time, and independently of Descartes, the French lawyer-mathematician Pierre de Fermat (1601–65) also created coordinate geometry. By all historical accounts, Fermat was the superior mathematician. Characteristically, however, Descartes snubbed him, proclaiming that "our nephews will find nothing in this material [Fermat's analytical geometry] that I could not have found as easily as they, if I had wished to bother looking." Thus he had no "wish to see the demonstration of Monsieur de Fermat." Descartes's penchant for self-promotion and his disdain for intellectual rivals have contributed to the engorgement of his historical persona, usually at others' expense. Nevertheless, Descartes did go far beyond Fermat and his other contemporaries, and he did distinguish himself by integrating the new mathematical insights and techniques into a profound and robust philosophical vision. Lest we forget, he was very, very smart.

Whereas Viète had unsuccessfully attempted the union of algebra and geometry with a correspondence between arithmetic and geometric proportions, Descartes took another tack, claiming that Viète's "whole nomenclature" for geometric magnitude "must be abandoned." Rather, magnitude of whatever sort or power, even a "cube or biquadratic, ought never be presented to the imagination in any form other than that of a line or a plane." How could this be accomplished? How could magnitudes or dimensions of higher powers (x^2, x^3, x^4 . . .) be presented to the imagination using only lines and planes? Descartes's answer, succinctly stated, was by correlating any point on a line or plane uniquely with a pair of numbers (and ultimately with a single number, since any number can be understood as a relation or ratio of two numbers; thus 2 can be expressed as the ratios 2:1, 6:3, 12:6, or as the fractions $^2/_1$, $^6/_3$, $^{12}/_6$, and so on). With this proposition, he made explicit a one-to-one correspondence between numbers and points and posited the relations between them as directly commensurable.

In setting up his problems, Descartes designated a straight line to serve as the reference for all other lines, curved or straight, that he wished to examine. On the reference line he chose a point "at which to begin the investigation." From there he proceeded to determine the relations between various lines and figures on the premises that "all points of those curves which we may call 'geometric', that is, those which admit of precise and exact measurement, must bear a definite relation to all points of a straight line, and that this relation must be expressed by means of a single equation."

Unfortunately there is no royal road to learning in these matters, but perhaps we may render Descartes's reasoning somewhat more tractable with an exposition of his analysis drawn from our own times. His central claim is that an ordered pair of numbers can serve as the coordinates of a geometric point. To see this, we can picture the coordinates as lying on a rectangular grid laid out in horizontal and vertical lines. Then we select one line to serve as the base horizontal line and one to serve as the base vertical line. These are the axes of the coordinate system. (They may be imagined as a street map, with an east-west "Main Street" cut perpendicularly by a north-south "Central Avenue.") The re-

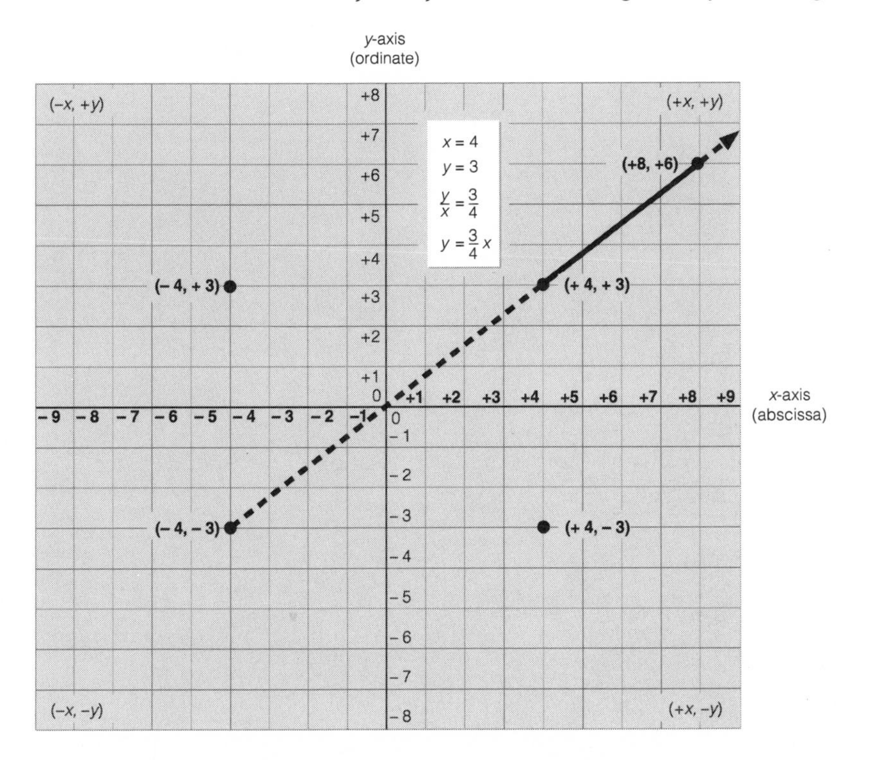

FIGURE 5.2. A Cartesian Coordinate System

maining lines (or streets and avenues) are drawn in at regular intervals (blocks) parallel to each axis and the intervals are noted by numbers (see figure 5.2.).

To locate any point on this map we simply count the number of blocks or intervals from the axes. In figure 5.2, for example, x is the horizontal axis, y the vertical axis, and a pair of numbers (4,3) identifies a unique point, 4 blocks to the right (of y) on the x-axis and 3 up (from x) on the y-axis. Ever since Gottfried Wilhelm von Leibniz (1646–1716) introduced the terms, a line parallel to the x-axis has been called the "abscissa" and one paralleling the y-axis the "ordinate"; in stipulating an ordered pair, the abscissa is conventionally given first, as in (4,3), $x = 4$, $y = 3$. Because the axes divide one another, each axis is considered 0, and their intersection (0,0), a point called the "origin." Thus (4,0) represents a point at four intervals to the right of 0 on the x-axis and lying on the x-axis itself.

Moreover, on the horizontal x-axis, to the right of the origin, the natural sequence of intervals increases positively (1, 2, 3, . . .), while to the left of it, the sequence is negative (−1, −2, −3, . . .). Similarly with the y-axis: Above the origin lie positive intervals, below it negative. Of course, the scale and size of intervals are arbitrary and depend on the purposes and applications at hand. Negative numbers in any pair are mapped accordingly, as shown by the ordered pairs (−4,3), (−4,−3), (4,−3) in figure 5.2. (We should remember that Descartes himself rarely used the now familiar rectangular coordinates described here; rather he constructed his equations with reference to oblique coordinate lines. Also, neither he nor Fermat admitted negative values for the coordinates; these were later additions.) In sum, the layout of the coordinate system comprises a grid with four separate quadrants, on which any geometric point may be uniquely plotted by an ordered pair of numbers, positive or negative. In such a system the one-to-one correlations between number pairs and points reveal a new, profound interpretation of the cardinal principle of number, uniting algebra and geometry.

If points can be mapped on a coordinate system, then so too can lines and geometric figures. A straight line simply connects two points, each of which is designated by an ordered pair of numbers, say (4,3) and (8,6), also shown in figure 5.2. Once having drawn the line through these points, and having extended it, we could locate other points along it as well. However, already in the first two points we have enough numerical information to establish an equation for the line. Since $x = 4$ and $y = 3$, dividing equals by equals gives us $y/x = 3/4$ (an equation of ratios, which says "y is to x as 3 is to 4"). Multiplying both sides of the equation by x produces $y = 3/4x$.

This is the algebraic equation of the line. It says that for every interval the line proceeds to the right of the y-axis on the abscissa, it rises ¾ of an interval above the x-axis along the ordinate. Similarly, the equation $y = x$ means that a line passes through one interval of y for every interval of x; thus $y = x$ represents a diagonal line bisecting the right angle made by the intersection of y and x, a 45 degree slope. In brief, any straight line can be expressed by a linear equation representing the coordinate points that lie on it.

For Descartes the real application of his insight lay less with straight lines than with curved ones, which would enable him to map the movements of material bodies with precision. Devising the equation for a circle builds on the foregoing with additional steps. Let the origin point (0,0) serve as the circle's center. The Euclid in all of us knows a circumference is a curved line everywhere equidistant from the center. Assume this distance, the radius, is five intervals or units. Now, any point on the circumference can be expressed as an ordered pair of coordinates. So the question becomes what properties or relations distinguish these ordered pairs from other points. If we take a point P on

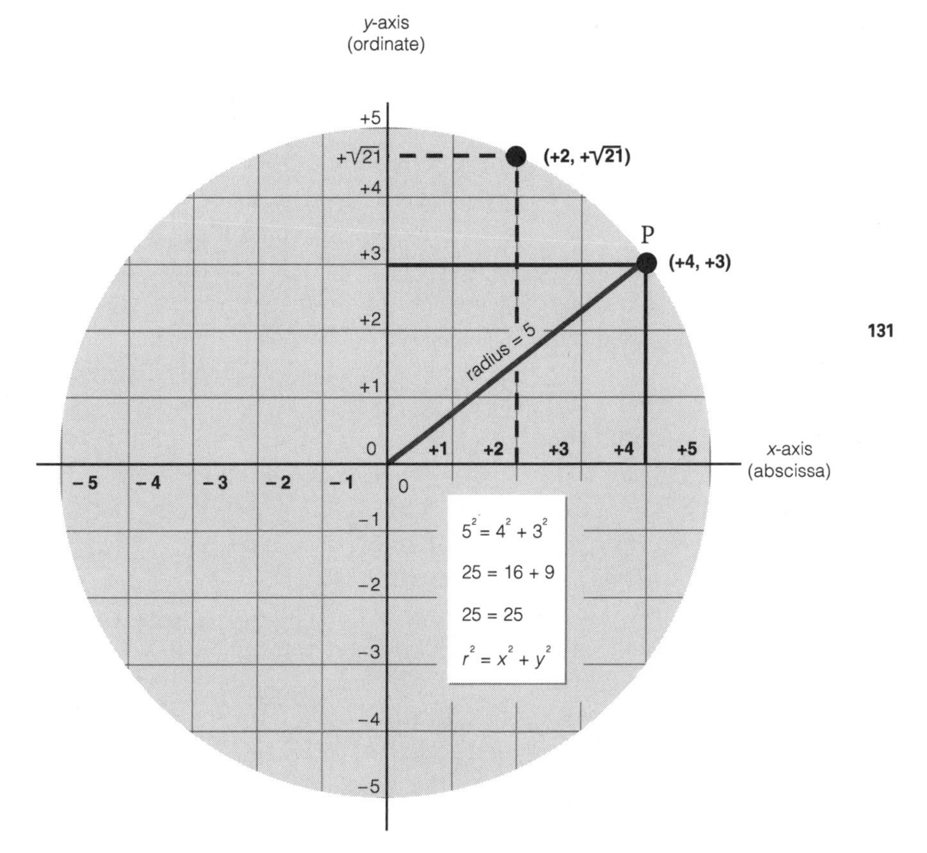

FIGURE 5.3. The Equation of a Circle

the circumference, as in figure 5.3, we see its coordinates are x and y. We see, further, that x and y form the sides of a right triangle whose hypotenuse is the radius. From Pythagoras, the square of the hypotenuse equals the sum of the sides' squares. Hence, given the radius as 5 we can form the equation $5^2 = x^2 + y^2$, using the notation introduced in the seventeenth century. This is the equation of a circle whose radius is five intervals. It must be satisfied by all points on the circumference, and it may be algebraically manipulated to find any one of them. For example, say we let $x = 4$ and we want to discover the value for y; then $5^2 = 4^2 + y^2$, $25 = 16 + y^2$, $y^2 = 25 - 16$, $y^2 = 9$, $y = \sqrt{9}$, $y = 3$; the coordinates (4,3) satisfy the equation and give us a unique point on the circumference.

We can find out another bit of information about coordinates by letting $x =$ 2 in the same equation. In this case $5^2 = 2^2 + y^2$, $25 = 4 + y^2$, $y^2 = 25 - 4$, $y^2 =$

21, $y = \sqrt{21}$. The $\sqrt{21}$ is an irrational number; it cannot be expressed as a rational ratio between two integers. Yet the ordered pair $(2,\sqrt{21})$ satisfies the equation and represents a unique point on the circumference. Both rational and irrational numbers, then, can be correlated uniquely in a one-to-one, cardinal correspondence with the coordinate points of the grid. Together the rational and irrational numbers make up the domain of real numbers.

It follows that the axes represent arithmetic continua of real numbers, the ordinal or natural sequence of increasing magnitude. Mathematicians today describe continua as infinitely dense. This means that between any two numbers on a continuum one can always add in or interpolate another number, or that a continuum may be extrapolated (or extended) to infinity. Density permits mapping an irrational number (like $\sqrt{2}$) as a unique point on a continuous line like our y-axis. Rigorous proof of the correlation between real numbers and points on a line was made only with the work of late-nineteenth-century mathematicians Georg Cantor and Richard Dedekind. Nonetheless, both Descartes's and Fermat's geometry assumed these correlations and the density of continua—the cardinal and ordinal principles of number.

The foregoing illustrations used numbers in the equations of lines and circles. But following Viète's lead, Descartes, Fermat, and others adopted the practice of generalizing from numbers with letters. Any line can be expressed by a first-degree (or linear) algebraic equation of the form $ax + by = 0$, where x and y are the coordinates and a and b the intervals on the abscissa and ordinate (or, simply, coefficients of x and y). The equation of any circle is the second-degree (quadratic) equation $x^2 + y^2 = r^2$, where x and y designate coordinates of any point on the circumference of a circle, and r its radius. Other, general quadratic equations can represent all the curves found in cross sections of the cone (circle, ellipse, hyperbola, and parabola), whose study had been revived in the late sixteenth century. Exploring properties of conic curves in any depth had proven intractable to Euclidian methods of construction with ruler and compass. With Descartes's analytical scalpel they could now be dissected and examined with symbolic equations.

In fact, most of Descartes's mathematical achievement lay in rendering the geometry of Euclid and other Greeks into algebraic form; he himself did not generate many new curves from newly created algebraic equations, as did Fermat. For our purposes we need not delve any further into the techniques of coordinate geometry but only highlight their central idea. Whether plotting new curves from given equations or finding new equations to express known curves, these techniques enabled one to move back and forth between the worlds of space and algebraicized number. The cardinal correlations between points and numbers united the ordinally arranged continua of geometry and arithmetic into a single, abstract science, universal mathematics.

A New Way of Informing

The significance of the new analytic techniques devised by Descartes and Fermat can scarcely be overdrawn. Mathematically they were revolutionary, establishing—in the assessment of mathematics historian Edna Kramer—"the very core of [modern] mathematics." Once created, they were extended and generalized step by step to produce a most powerful means of expressing all different kinds of abstract relations and of drawing inferences from them. From the perspective of our narrative, the techniques were equally revolutionary in their capacity for expressing information in symbolized form, and hence for ushering in the modern information age. To reiterate the claim even more bluntly: Modern, relational mathematics would become the first, fully symbolized and abstract language capable of storing, organizing, and manipulating information.

Descartes and his contemporaries did not describe this development in quite the same way we have chosen. They used different terminology, one still reverberating with scholastic overtones and meanings. And in our haste to see the dawning of a new age, we need to beware historical anachronism. With this caveat, we may nonetheless understand Descartes's achievement through the words of Norbert Wiener (one of the gurus of our own information age): "Information is a name for the content of what is exchanged with the outer world as we adjust to it, and make our adjustment felt upon it." In this general, descriptive sense, 'information' applies exactly to the Cartesian project of regrounding and reunifying knowledge. With his techniques of coordinate geometry, science of universal mathematics, and philosophical exposition of clear and distinct ideas, Descartes formulated a wholly novel way of shaping the contents of our exchanges with the outer world, a new technology for managing information. This was the analytical vision, whose means of storing and processing information were found in the mathematical language of abstract operational symbols.

Once we recognize mathematics as an abstract information technology, enabling us to interact with the world, its contrast with literacy-based forms of information becomes all the more striking. Cartesian points and equations do not store "things" abstracted from experience and represented by the nouns and adjectives that make up taxonomies. Nor do they store groups or collections of things in the manner of early counting systems. Descartes conceived abstractions differently, starting with what he called the intuition of "pure and simple natures," which (while somewhat ambiguous) refer to the relations between things, rather than to things themselves. His points and equations express the abstract relations, which constitute the new form of information.

Furthermore, the new kind of information captures the relational features of natural phenomena. This was the practice begun by Descartes's slightly older contemporary Galileo Galilei (1564–1642), among others. Galileo did not have the precise analytical tools of coordinate geometry at his disposal; he never used algebra or even decimal fractions, relying instead on older, geometric expressions. Yet he had worked out something very similar in his "thought problems" and experiments. (In fact, he is often cited as the first scientist to use the modern scientific method.) Galileo named his method "resolutive-compositive," by which he meant that it resolved (or analyzed) complex phenomena into their constituent components, then composed them in their correct, mathematical relations (or formulas).

Among his many investigations, Galileo applied this method to the "compounded motion" of "moveables" (including projectiles such as cannonballs), a motion he resolved into its horizontal and vertical components. He then considered each component to be, in effect, the axis of a coordinate system, with one axis standing for intervals of time, another for units of distance. The compound motion of the body's actual path was represented by a curved line between the axes, and was expressed in a formula or ratio that related the two coordinates exactly. The points on the projectile's path gave it unique coordinates while the formula of the line equated the different rates of speed (whether constant or accelerating) designated by the axes of time and distance.

In Galileo's analysis a point or number does not represent any particular object itself, but only a body's position at a given time—its momentary, temporal and spatial position relative to other temporal and spatial positions—a highly abstract conception. "Unpacking" the relational information contained in a point or equation will require, henceforth, devising the formula that captures the relations actually existing between phenomena. Done correctly and confirmed by experiment or exact observation, the simulations provided by formulas can be used to describe and map actual events in the natural world.

A coordinate system of analysis can store unlimited amounts of information because it can be set up and exploited in unlimited ways, depending on needs and interests. Coordinate axes will come to stand for the flow of time, intervals of distance, configurations of spaces, amounts of force (gravity, inertia, gas pressure, magnetism, electricity, and so on), the longitudes and latitudes of navigation, credits and debits, economic supply and demand—the list grows as we jump ahead of ourselves, but it is hard to avoid. Once the link has been made between phenomena and mapping their abstract relations on a coordinate grid, the yellow brick road into the land of analysis lies before us.

Finally, the points or equations hold what we can call fixed or invariant information about phenomena. Invariance means here that the relations between the *x* and *y* coordinates, cast in formulas, are either unchanging or mutually dependent, so that if one value (*x* or *y*) is transformed by algebraic techniques, the other is also, and in exactly the same way. Galileo's formulas thus will map invariantly the motion of all bodies, including cannonballs, moving under similar circumstances. The invariance in these relations enables us to predict changeable phenomena in a manner we can characterize nontechnically as determined and linear. The determinism of predictions follows because the equations of the axes are exact and therefore can produce, or predict precisely, any given points of the line itself.

The linearity arises from the imagination's perception of the continuous "line" of motion that describes whatever body or particle is being graphed. Descartes and others after him often referred to geometric figures in this manner as present to the "imagination." The imagination becomes a visual halfway house on the road to abstraction; one can "see" moving bodies and geometrized spaces on paper. Lying completely beyond sight or image, however, the fullness of conceptual abstraction is achieved only with the mathematical equations (of continuous functions) that designate the lines, not with compass and ruler.

As the seventeenth century progressed and as analytical techniques were refined and developed, the invariance and predictability of abstract relations came to inspire in many learned men a growing confidence in the symbolic language of mathematical analysis. Not only did it identify and extract new and reliable forms of information from experience, but even more important, it became the standard enabling one to sort out information and to eliminate "false" information from the catalog. In contrast to the sterile classifying terms of natural language (pilloried by Molière's "opium causes sleep because of its dormitive powers"), this novel instrument of information mastery seemed to promise results, particularly in the investigations of physics and astronomy, more generally in the study of nature writ large. To paraphrase Galileo, although our knowledge might not tell us how to go to heaven, knowing predictably how the heavens go gives us a reliable compass for uncharted seas of information.

The Reintegration of Knowledge

The foundations of modern numeracy—those principles of cardinal correlation and ordinal recurrence—had long been present in early quantitative thinking but were transformed dramatically during the sixteenth and seventeenth centuries into a purely symbolical language of immeasurably greater power and reach. A positional counting system, symbolic notation for mathematical operations, and a completely symbolized algebra had prepared the way for the Cartesian revolution in thought. Born of a dream, the revolution swept away all before it—all the lists, classes, categories, genera, species, commonplaces, and taxonomies of the classical age—and replaced them with a new information idiom, the analytical vision of knowledge.

Well, not quite. Philosophers always consolidate and organize what scientists discover—at least that has been the view since Aristotle's opening salvo in the *Posterior Analytics:* "All teaching and all intellectual learning come about from already existing knowledge." Donning his philosopher's cap, Descartes too sought consolidation and organization, the integration of new learning with old. Even as he sliced away scholastic classifications and humanist commonplaces with his razor of systematic doubt, he fervently held that the new knowledge would be thoroughly subsumed under a proper appreciation of substantive reality. Like most of his contemporaries, he continued to cling doggedly to features of classificatory thinking, primarily the categorical substances of mind and matter, even while he proffered mathematics as the new basis for clear and distinct reasoning.

This belief jars the "modern" in us as we focus on the revolutionary nature of Descartes's innovations. We need to bear in mind that his achievement occurred in an intellectual world strikingly different from our own. Descartes remained convinced that with his *mathesis universalis* someone (he himself) had finally gotten the world's substances right, that the formulas conceived by the mind described the real world of extended matter, and that mathematics and substance were part and parcel of the universal order of wisdom, *sapientia* or *sagesse.* He sometimes wrote of "primitive notions," substance and number, which provided the "models" for all we could know. And he often expressed his profound belief through a favorite metaphor, the tree of knowledge. Unlike its Biblical counterpart, which gave us only good and evil, his tree promised more, "philosophy as a whole." It was a tree "whose roots are metaphysics, whose trunk is physics, and whose branches, which issue from this trunk, are all the other sciences," moral as well as natural.

Even so, Descartes's desire to reintegrate knowledge and wisdom harbored a profound ambivalence, a tension between the techniques of his analysis and his conviction as to the reality of substances. To manage this ambivalence Descartes reinvented God, or at least a new argument for God's existence. We have, he reasoned, some clear and distinct mathematical ideas that tell us something about the world, that frame our information exchanges with it. We know the information is true, because a perfectly good God could not deceive us into thinking otherwise. And we know the good God exists because our clear and distinct ideas exist. In this "Cartesian circle," God guarantees the "content of what is exchanged with the outer world." The need for such an information guarantor, a deus ex machina, reveals a fundamental ambivalence about analytical abstraction, which for Descartes still had to be linked to the substantive world of God's creation.

Seventeenth-century wit and letter writer Madame Sévigné once commented that, like coffee, Cartesianism in the end would only be a passing fashion. Obviously, with this double absence of insight the muse was not at the moment guiding her pen. Long after Descartes, in fact, other learned men would continue grappling with the ambivalence that had plagued him, the theoretical problem of unifying mathematics and classification. Philosophical debates about the Cartesian mind-body dualism even persist into our own era. At the same time, in the more practical realm of knowledge organization, Descartes laid the groundwork for an extensive restructuring of the sciences and humanities. This practical restructuring was to foster another line of development, one in which the analytical vision would imperiously seek to subsume the vestiges of substance under its own ordering of experience.

The eighteenth-century *philosophes,* Diderot and d'Alembert, would begin with the same metaphor that Descartes had used, the universal tree of knowledge, as their inspiration for a practical encyclopedia of organized knowledge based on the analytical vision. But while Descartes intended the metaphor to depict both wisdom of the ancients and the knowledge of the moderns in a grand, synthetic reunion of the two cultures, the encyclopedists would reach beyond the Cartesian "spirit" of a philosophical system to promote a "systematic spirit" of information gathering and organization. Extending the analytical vision would lead them to a new metaphor: a world map of mathematical abstractions, which would become the acme of the modern information age.

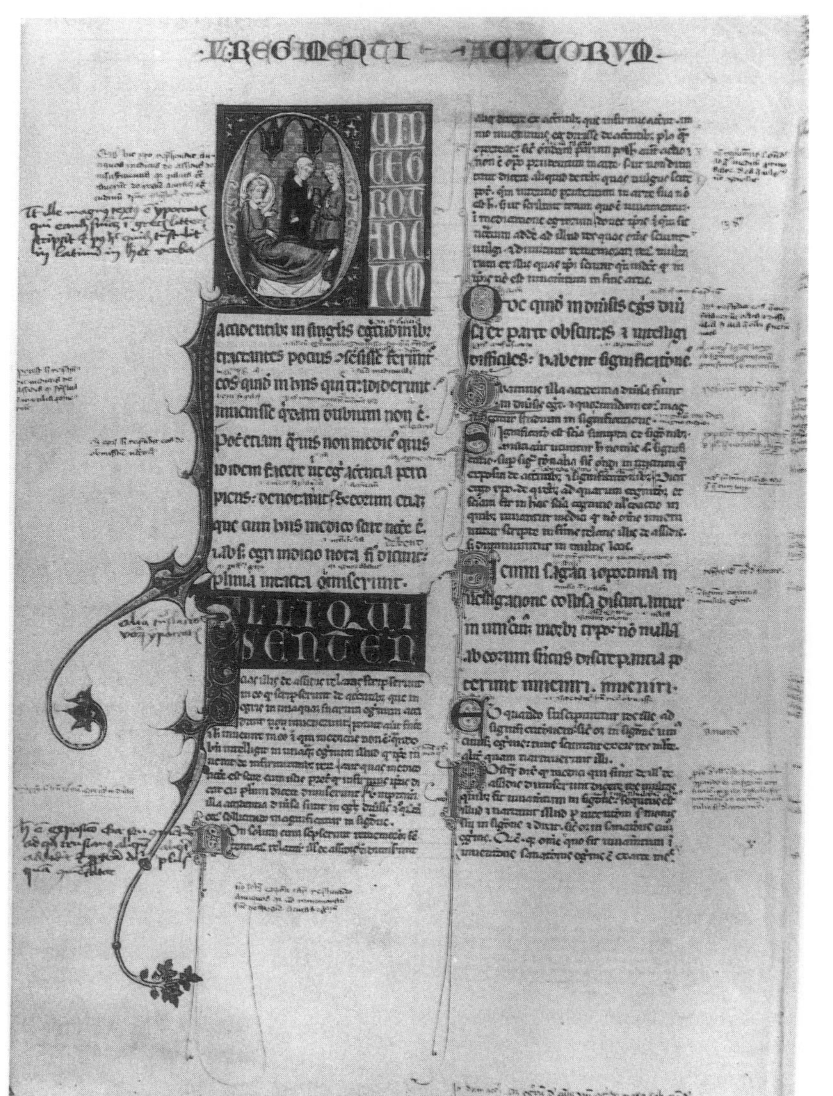

PLATE 7. *A glossed manuscript.* This late-thirteenth-century Latin translation of a medical work by the Greek physician Hippocrates combines text (in the large book hand) and commentary (in the small book hand). Marginal and interlinear notes in several cursive hands constitute additional layers of commentary. The miniature in the letter at the upper left of the page depicts two doctors discussing a patient. Colorful illustrations and ornate letters served as visual hooks for remembering the contents of the manuscript. From the Österreichische Nationalbibliothek, Vienna.

Librum ihesu filii sirach dicit se ieronimus apud hebreos repperisse non ecclesiasticum ut apud latinos sed parabolas prenotatum. cui iuncti erant ecclesiastes et canticum canticorum ut salomonem non modo librorum numero sed et materiarum genere coequaret. librum vero sapientie non repperisse apud eos sed magis grecam redolere eloquentiam. quem nonnulli philonis esse affirmant. Ecclesiasticus vero sicut et ecclesiastes. ecclesie utilimus est qui congregator vel collector interpretatur. sicut ille concionator sed ecclesiastes ad christum refertur et ad quemlibet predicatorem. ecclesiasticus qui propter excellentiam virtutum suarum panarethos i. omnium virtutum capax appellatur. cuius tanta claritas tantaque latinitas est ut ipse sibi commentarius sit.

Omnis sapientia a domino deo est et cum illo fuit semper et est ante evum. Harenam maris et pluvie guttas et dies seculi quis dinumeravit? Altitudinem celi et latitudinem terre et profundum abyssi quis dimensus est? Sapientiam dei precedentem omnia quis investigavit? Prior enim omnium creata est sapientia et intellectus prudentie ab evo. Fons sapientie verbum dei in excelsis. et ingressus illius mandata eterna. Radix sapientie cui revelata est et astucias illius quis agnovit? Disciplina sapientie cui revelata est et manifestata et multiplicationem ingressus illius quis intellexit? Unus est altissimus

Omnis sapientia a domino deo est etc. Incipit ab eterna dei sapientia que christus est. cum patre est semper ante secula secundum illud. In principio erat verbum et verbum erat apud deum etc.

Harenam maris etc. Corporalia spiritualibus comparat ut ex difficillimis que omnino impossibile est estimentur. Sicut enim hec in certo numero et mensura et pondere condita sunt nemo potest dinumerare. quanto minus sapientiam dei sine initio et sine fine inenarrabilem inestimabilem poterit investigare.

Creata secundum humanitatem in consilio dei patris. scilicet pro salute humani generis predestinata est incarnari. Notandum autem quod propter unitatem persone aliquando christus dicitur genitus aliquando creatus.

A domino deo etc. Quia christus fons vite lux vera de patre nascitur et omnia per ipsum facta sunt et sine ipso factum est nichil. Quisquis hac sapientia caret in tenebris ambulat et nescit quo vadat. Sapientia enim huius mundi stulticia est apud deum et prudentia carnis mors.

Fons sapientie. Origo omnis sapientie a verbo dei procedit. quia deus ex deo semper cum patre manet in celis. sed ingressus in mundum per dispensationem humanitatis dedit mandata salutis. unde verba vite eterne habes.

Radix ostendit neminem posse profunditatem dei penetrare vel magnitudinem bonitatis eius et misericordie. unde ad liberationem et illuminationem humani generis homo fieri dignatus est. unde generationem eius quis enarrabit?

PLATE 8. *Text and commentary as visual pattern.* In this thirteenth-century gloss of the book of Ecclesiastes, not only the ornamental letters but also the very layout of text and commentary serve to imprint the contents of the page in memory. Note that plenty of room is left for interlinear and marginal glosses in a small, cursive script. From the Houghton Library, Harvard University.

PLATE 9. *The house of grammar* (woodcut attributed to Voghterr, 1548). The different stories and rooms of the house represent different topics and subtopics in the study of grammar. From Lina Bolzoni, *La stanza della memoria* (Turin: Giulio Einaudi, 1995), p. 261.

PLATE 10. *A city as a memory system,* from Johannes Romberch, *Congestorium artificiose memorie* (Venice, 1533). Here the buildings of a city—such as the bookstore (*bibliopola*), barbershop (*barbitonsor*), and slaughterhouse (*bovicida*)—serve as memory structures, with each edifice storing a different kind of information. From Bolzoni, *La stanza della memoria,* p. 260.

PLATE 11. *"Grammar" personified as a memory image,* from Johannes Romberch, *Congestorium artificiose memorie* (Venice, 1533). Every aspect of this image is pressed into mnemonic service, with Grammar's stance, the ladder, tools, and birds all serving as visual hooks for the recollection of information. Elsewhere in his treatise, Romberch shows how to devise visual alphabets for storing information by alphabetic means, using the names of different birds and the shapes of various tools to recall letters. From Frances A. Yates, *The Art of Memory* (Chicago: University of Chicago Press, 1966), p. 113.

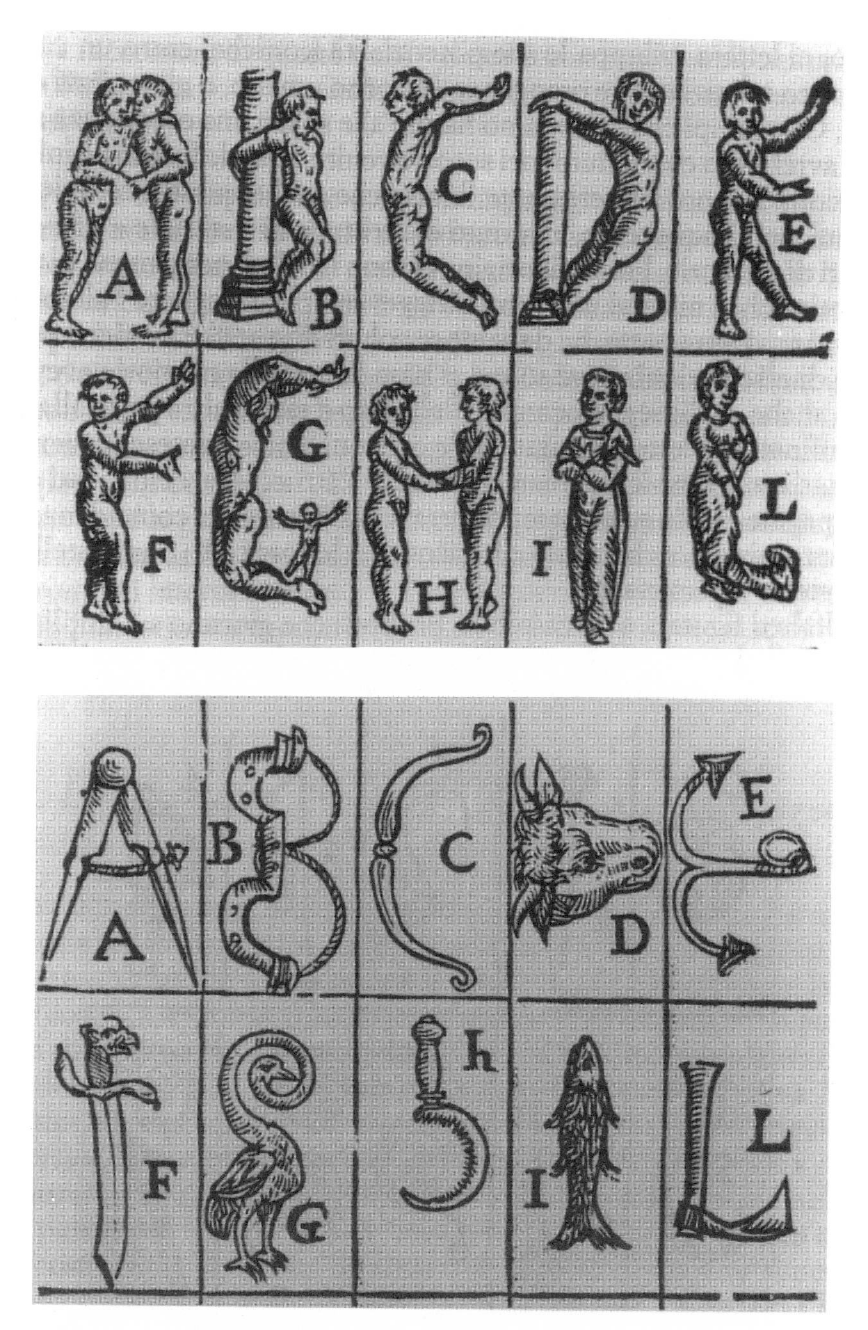

PLATE 12. *Visual alphabets,* from Giovan Battista Della Porta, *Ars reminiscendi* (Naples, 1602). The objects, animals, and human figures provide a vivid means of storing and recalling alphabetically arranged information. Memory can be built up in layers by affixing these alphabetic devices to an allegorical image, situating the image in a room, the room in a house, and the house in a city. From Bolzoni, *La stanza della memoria,* p. 101.

PLATE 13. *A contest between abacists and algorithmists,* woodcut dated 1508. Personified as a woman, the spirit of mathematics presides over the competition. Until well into the sixteenth century, speed and facility of calculating enabled the abacists to hold their own against the algorithmists, who wrote their calculations in the new Arabic numerals. From Frank J. Swetz, *Capitalism and Arithmetic: The "New Math" of the 15th Century* (La Salle, Ill.: Open Court, 1987), p. 31.

PLATE 14. *A Pascaline.* One of the first calculating machines, it was designed and built by French philosopher Blaise Pascal in the 1640s. The device could both add and subtract, with subtraction being accomplished by a method known as "adding complements." From the Centre Pascal, Clermont-Ferrand.

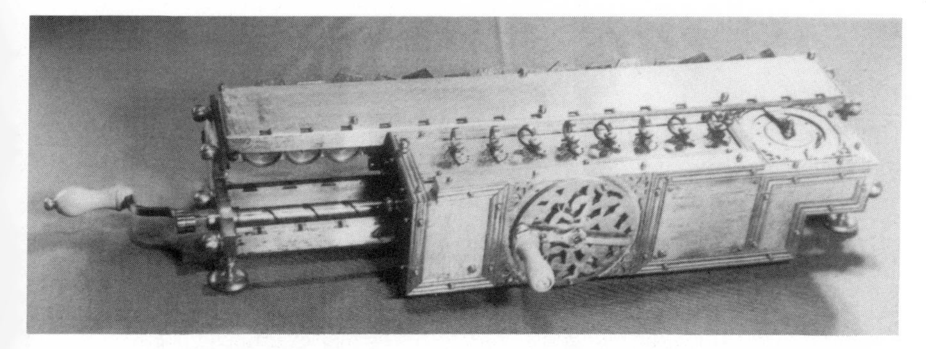

PLATE 15. *Leibniz's calculator.* A reconstruction from Leibniz's design, known as the "Stepped Recknoner." More advanced than the Pascaline, this machine used a special gear, now called the "Leibniz wheel," which served as a mechanical multiplier. From the Deutsches Museum.

CHAPTER SIX

The Analytical World Map

> The information of mathematics is very considerable, and even in a way inexhaustible. —D'ALEMBERT, *Preliminary Discourse*

The Black Box

The emperor Napoleon once complimented his celebrated and vainglorious countryman, astronomer-mathematician Pierre-Simon Laplace (1749–1827), on the appearance of a volume in the latter's *Treatise on Celestial Mechanics.* No mean mathematical talent himself, at age sixteen Napoleon had ranked forty-second on a national listing of mathematics students, placed there by his examiner at military school, the very same Laplace. Yet in later life, ever the emperor, Napoleon could not resist chiding his former examiner for having "written this huge book on the system of the world without once mentioning the author of the universe." To which the eminent scientist replied, simply, "Sire, I had no need of that hypothesis."

The truth of this famous, perhaps even apocryphal, anecdote in the history of science grows with each retelling. It does so because Laplace's incisive riposte captures succinctly the intellectual ground shift from the seventeenth-century's religious concerns, with its longing for *sapientia,* to the eighteenth-century's secular age of Enlightenment, with its voracious appetite for human knowledge. We remember Laplace's quip, seldom that of his fellow mathematician, Lagrange, to whom Napoleon later related the incident and who responded in a much different vein: "Ah, but that is a fine hypothesis. It explains so many things."

Descartes had died in frosty Sweden believing that the techniques of coordinate analysis and their underlying universal mathematics would provide the means for reintegrating knowledge and wisdom. Within two or three generations, he imagined, the natural and human sciences would be unified into a single, all embracing philosophical system—if only others would follow his lead. Yet by the time of Laplace's rebuff to the emperor, the analytical vision had taken a much different turn. On the one hand it had become more rigorous, refined, and narrowly focused in its methods. On the other hand, it was at the same time growing more expansive and imperial with each passing year, bringing more and more bits of knowledge into its pale, even while leaving wisdom beyond the pale.

Behind this twofold trajectory in the analytical vision lay the rapid growth of modern, relational mathematics, which soon had far surpassed its origins in Cartesian coordinate geometry. The most profound moment in the evolution of analysis occurred during the late seventeenth century with the dual, independent discoveries of the calculus by Isaac Newton (1642–1727) and Gottfried Wilhelm von Leibniz (1646–1716). Throughout the following century, sustained developments refined this powerful tool into an extensive, logical array of abstract symbols, a new kind of language that has come to occupy for many a critical location between us and our surroundings. With it one can devise and sustain a manageable structure to frame information, the content of our exchanges with the outer world. "God made the integers," wrote mathematician Leopold Kronecker after the Enlightenment, but still conveying its deistic spirit, halfway between Laplace and Lagrange, "all the rest is the work of man."

To detractors and devotees alike, this new language became steadily more mysterious and remote, a black box of symbols, operations, and techniques that according to one's perspective either performed marvelous tasks or murdered to dissect. For the arithmephobe, or (more benignly) the innumerate, the black box stood suspiciously outside the purview of normal experience, a haunted house wherein dwelt the "ghosts of departed quantities" according to the divine Bishop Berkeley. (In this vein it still amuses, and comforts somewhat, to recall Bertrand Russell's definition of mathematics as "the subject in which we never know what we are talking about, nor whether what we are saying is true.")

For its enthusiasts, the analytical vision encouraged many savants to extend its black box operations all the further, to recast all learning, moral as well as scientific, with the techniques and methods provided by mathematics and its accompanying experimental method. In the narrow sense, organizing knowledge with the tools of mathematical abstraction entailed reducing complex phenomena to their constitutive components, properly described with their respective mathematical formula. This was the method Galileo had first introduced as "resolutive-compositive" and whose chief instrument Descartes and Fermat had provided with their analytical geometry.

In its broader dimension, the reductionism inherent in analytic techniques came to be seen as the answer for the problem of too much information, that "horrible mass of books which keeps on growing." (Thus grumbled Leibniz, among others.) The onslaught of new information begun in the sixteenth and seventeenth centuries continued unabated into the eighteenth. The only possible way to master it, many came to believe, lay in reducing it to some analytical order, which allowed one to create a new and different hierarchy of knowledge. Only with a radically different hierarchical arrangement, derived from mathematical reduction, could information be organized in a way that avoided all the pitfalls of classifying.

During the mid-eighteenth century, the attempt to create such a new reductionist hierarchy surfaced most clearly in the awe-inspiring labors of Denis Diderot (1713–84) and Jean Le Rond d'Alembert (1717–83) with their editorship of the *Encyclopedia* (1751–72). Amid the ideologically charged, political turbulence surrounding publication of the *Encyclopedia*, a dominant intellectual objective steadfastly guided the editors' project. Both men sought to articulate the analytical vision and to derive from it a theoretical and practical ordering, an "enchainment of knowledge" (Diderot's phrase), an encyclopedia bound together with mathematical "chains of reasons."

Both editors believed that the epistemological canons of the new mathematical sciences could produce and justify with equal validity a "detailed system of human knowledge," in effect a mathematized taxonomy. Mathematics would ground a novel structured hierarchy of information. And were the hierarchy extended far enough, it would bring to fruition Descartes's dream of unifying the natural and moral sciences, knowledge and wisdom, whose metaphorical embodiment—the tree of knowledge—the editors embraced as the inspiration and frontispiece of their great work.

Although inspired by the analytical vision of the Cartesian project, d'Alembert and Diderot approached it from quite the opposite direction than had their predecessor. Forget the "spirit of the system," wrote d'Alembert in reference to Descartes's and the seventeenth century's penchant for wanting to know it all; stick to the true "systematic spirit" (his idiom for what nowadays some philosophers call instrumental reason). The new and powerful techniques of analysis associated with the calculus and with the experimental scientific method would piecemeal net and arrange all that could be known with any assurance. The remainder could then be consigned to the dust bin of superstition and religious enthusiasm, the vestigial remnants of humankind's infancy.

To depict their new stratagem for systematically extending the reach of analysis, the editors grafted another metaphor onto the tree of knowledge. This was the "world map" (*mappemonde*) of cognitions, which figuratively situated and arranged all the natural and human sciences, from the theoretical to the practical and liberal arts. The world map served as a general guide for the "great regions" of knowledge. By contrast, the "particular" or "individual maps," which comprised the subjects of the various encyclopedia articles, helped readers explore specific kingdoms, provinces, and countries of the cognitive globe. Not only did the world map guide one geographically, as a "Systematic Chart" it also structured the topography of knowledge, allowing one to see what "rank" any subject occupied, and in this fashion to understand its "place" in the overall "order and connection" of phenomena.

Growing out of the encyclopedia project of Diderot and d'Alembert, the world map came to signify something far more grandiose and general: the mathematical mapping or correspondence of mental objects with the "real," extra-mental referents of the world. Through this metaphor the reductionism inherent in mathematics would now provide the new hierarchy of knowledge demanded by order-seeking minds of the age. Thus the *Encyclopedia* stands as the first major attempt to create an entire compilation and organization of knowledge based on the abstractions of modern numeracy. For this reason we may consider it the culminating moment in the rise of the modern information age. By the end of the eighteenth century, analysis towered triumphantly over the ashes of taxonomy.

As the driving force of eighteenth-century, symbolic thinking, and the world map of knowledge, the calculus engineered this analytical triumph. Descartes's coordinate geometry had been a momentous first step in the development of modern numeracy. With its symbols and algorithms, the calculus set the gears of Cartesian analysis in motion, as it were, and provided the critical breakthrough into the highly abstract and reductionist realm of modern information.

Accordingly, before we turn to the encyclopedists and their world map, we must peer for a moment inside the black box and look at its symbolic language, its mechanism. We must look at the calculus. (Actually, even if the symbols of the calculus are ghosts, as Berkeley believed, they are like Casper, rather friendly once you get to know them.) This will not cure one's arithmephobia, but it will indicate historically how and why the symbolic language of analysis grew steadily more distant from natural language, so much so that by the eighteenth century much of the information contained in mathematical symbols could neither be comprehended nor articulated exactly in words.

In our glimpse inside the black box we shall confine our focus to the three core ideas of mapping, function, and limit, which together unify and help explain the procedures of the calculus. These ideas, and the techniques embodied in them, all grew historically and logically out of correspondence and recurrence, the hallmarks of numeracy. They reveal the technical means of eliciting the information potential from modern numeracy and, consequently, of making the analytical world map possible.

From Descartes to the Encyclopedia: *The Calculus and Classical Physics*

In the period we are discussing, the term 'calculus' simply meant calculation and included a variety of techniques and procedures one could follow in computing changes of all different sorts. New analytical tools were first devised to tackle specific problems in physics and astronomy during the seventeenth century. With their geometrical methods of treating fixed spaces and positions, the Greeks (Archimedes in particular) had developed the science of statics, the study of bodies at rest. By contrast, during the seventeenth century, concern lay with bodies in motion and with the calculations of force, distance, time, speed, acceleration, and other properties that pertained to dynamic or changing behavior of phenomena.

Newton called this motion "the flux" (from the Latin *fluere,* "to flow") and termed his techniques for calculating changes in motion "fluxions" or the "fluxionary method." Leibniz, too, devised techniques for calculating flow or motion rates and the "infinitesimal" (indefinitely small) changes in them. He developed his methods in two ways, as "differential" and "summatory" (later called integral) calculus, which have since become the two major branches of the discipline. The former calculates differences in rates of change, whereas the latter sums up or integrates change rates. For both men calculations meant following specific mechanistic procedures whose common term was and has remained, after Al-Khwarizmi, an algorithm.

In these procedures, the terms 'mapping' and 'function' closely intertwine one another around the cardinal concept of number. Mapping means correlating the members of different sets of objects so that there exists a one-to-one correspondence between the members of any two sets. Descartes had asserted that sets of real numbers and points correlated with one another, and that, accordingly, one could map points or numbers such as (x,y) onto a coordinate grid, and then devise equations to represent geometric figures. Thus in manipulating algebraic equations, one simultaneously mapped formulas onto space and vice versa. (Similarly, with a grid of latitude and longitude lines, one could navigate by following a map of geographical positions on global coordinates.)

Besides the mapping of (x,y) on a coordinate system, the use of x and y as abstract symbols emerged in the seventeenth century as a common practice growing out of Cartesian geometry. They came to be understood as abstract variables correlating in a one-to-one fashion with different physical features of nature, such as units of force, time, distance, mass, and the like. The actual correspondences were a matter of exact observation or experiment. The cardinal concept served to connect mathematical symbols to the world outside mathematics, and hence to map information about the outside world.

Although the term 'function' had been used occasionally before his own time, Leibniz recast it in its modern guise. Roughly, it means "depending on," and it refers to equations where the value of one variable (a number or a letter) depends on another variable and the changes the latter undergoes. In the equation of a line, $y = \frac{3}{4}x$ (our example from last chapter), y depends on x; it is the dependent variable, with x the independent variable. Any changes in x produce corresponding changes in y. That the variable y can be "mapped onto x" or is a "function of x" simply states in another way that the value of y depends uniquely on its one-to-one correspondence to x.

The Swiss mathematician Leonard Euler (1707–83; "analysis incarnate" to his contemporaries) formulated the conventional notation for a function as $y = f(x)$. Rhetorically this says "y equals the function of x" (or simply "the f of x"). For those unacquainted with the symbolic notation used in calculus, we should caution that $f(x)$ does not mean f times x in the manner of algebra. Rather, as a completely general symbol, $f(x)$ refers to any procedures or alterations x might undergo. It might be augmented $(x + a)$, diminished $(x - a)$, squared (x^2), multiplied by a factor or coefficient of $\frac{3}{4}$ (as above), or modified in vastly more complicated ways. (Similarly, the symbols Δx, dx, and ∂x, which we shall introduce later, do not designate the product of Δ, d, or ∂ times x, but stand for the procedural referents of x—difference, differential, and partial differential, respectively.)

The ideas of mapping and function, then, reveal how the methods of the calculus were devised to calculate and describe various changes in the value of x and correlate to them the dependent changes in the value of y. The great advantage of the calculus comes with the next step, with its ability to compare precisely not just changes themselves, but variations in rates of change. This is extremely important for the intuitive reason that few things change at a constant rate in natural and social flux; to the contrary, most rates of change are themselves changing or fluctuating (those "fluxions" again). In other words, change rates are variable.

Thus if we drive from home to a town a hundred miles away and arrive after traveling two hours, we have averaged fifty miles per hour. This is our constant, average rate of change (velocity in this instance) from home to town, surely good information to have if we intend to arrive at a specified moment. Yet our average rate of travel does not mirror very accurately the actual speeds at which we traveled. We certainly did not start our journey instantly traveling 50 mph; we accelerated and decelerated in traffic; we stopped and started for lights; the weather slowed us down; and so on. Among its many capabilities, the calculus can devise ways of finding these variations in rates of change at any given moment (or instant) and of correlating them to others. It can thus mirror more accurately our actual journey.

With the foregoing in mind, let us examine the ideas of mapping and function as they actually appear in a calculation. (We shall take up the notion of limit below.) Explicating the steps of a calculation will show just how the abstract symbols of the calculus came to stand for algorithmic procedures. Moreover, it will reveal the difficulty, even the impossibility, of describing in words the information stored and manipulated in the new analytical language. For the calculus would create a world of information that was quite literally unimaginable, beyond the net of our rhetorically rooted images. This was the world d'Alembert would call the "phantom" world—the world of Berkeley's ghosts.

To enter this *spiritus mundi,* we can continue our previous example of the coordinate grid and a straight line with the equation $y = \frac{3}{4}x$. Suppose we let the intervals along the x-axis correspond, one-to-one, with the advance of years, and those along the y-axis correlate with units of distance, measured in feet. The axes now represent two different types of change, and we can use the equation to compare them, say, to plot the growth of a tree's height over time. At the end of the first year it is ¾ of a foot tall; after two years its height is 1½ feet; after four years, 3 feet; and so forth. We see from our graph (figure 6.1) that the tree's average growth rate is ¾ of a foot per year. As we have already noted, the equation $y = \frac{3}{4}x$ states that for every interval x increases on the abscissa (x-axis, here meaning years), y increases ¾ of an interval on the ordinate (y-axis, feet).

Instead of comparing the whole axes, we can obtain the same result by comparing only segments of them. (Perhaps we found our tree after it had been growing a number of years and do not know the point of origin (0,0), when it

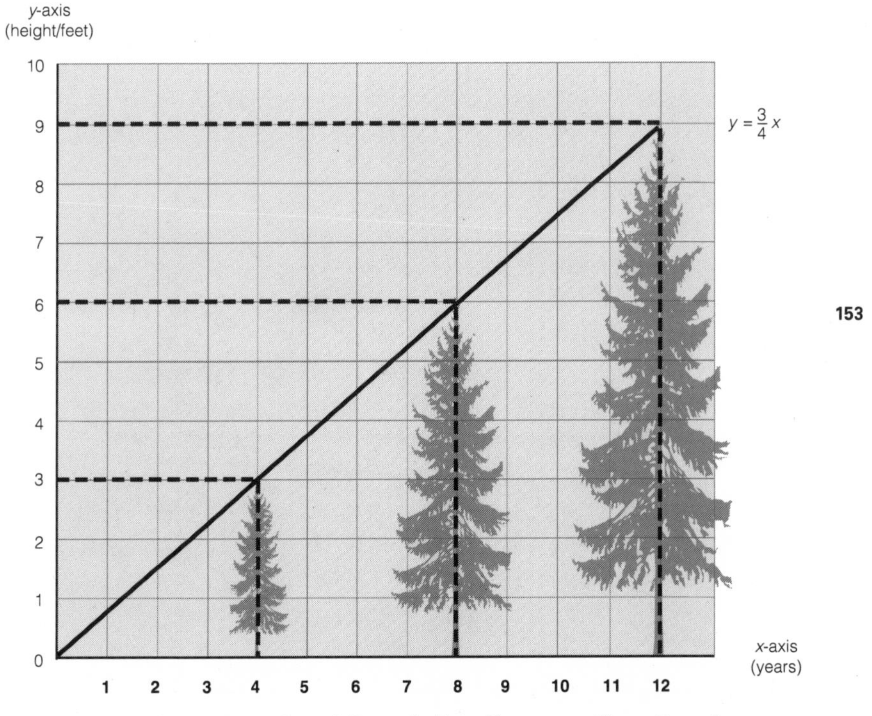

FIGURE 6.1. Average Growth Rate of a Tree, Shown as a Linear Equation

was planted.) Because the y-axis represents height, we may select a segment of it to represent a difference in height between two separate points, or the growth in height from one point to another. This difference is indicated by the symbol Δy as shown in figure 6.2. Likewise with the x-axis: the symbol Δx refers to a difference in time intervals. (The Greek letter delta (Δ) has long served as the conventional symbol for difference.) We watch the tree grow Δy feet taller in Δx years.

The growth ratio now will be the change in height divided by the change in years, $\Delta y/\Delta x$. In our example the height, Δy, is the difference between 6 feet and 3 feet, or $\Delta y = 3$; similarly the difference in years is indicated by $\Delta x = 8 - 4$, or $\Delta x = 4$. The new equation, $\Delta y/\Delta x = 3/4$, gives us the growth rate of the tree as the ratio between two constant rates of change, represented by the Δ segments of the y and x coordinates. Finally, multiplying both sides by Δx puts the ratio in the form of a function, $\Delta y = 3/4\Delta x$, which describes the change in the dependent variable Δy whenever there is a change in the independent variable Δx.

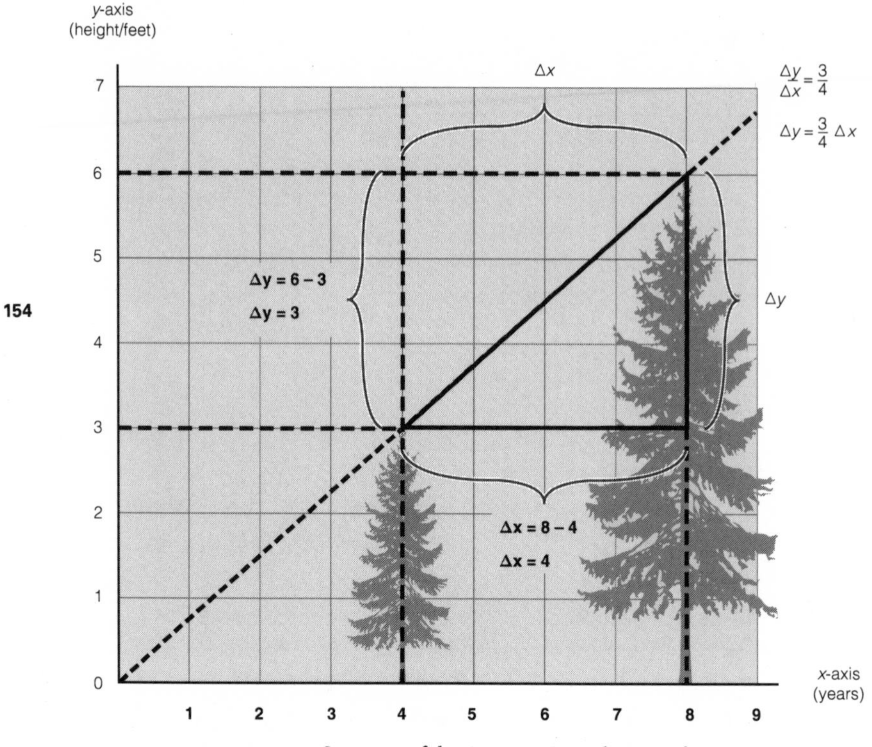

FIGURE 6.2. Segment of the Average Growth Rate of a Tree

The foregoing replicates a basic coordinate analysis, with only slight changes, and even seems cumbersome and unnecessary given our simple linear example, which remains quite intuitive, appealing visually to common sense. (It is, after all, a straight line.) Changes are seldom constant, however, and the power of the calculus begins to attain full fruition when comparing and calculating variations in different rates of change. As we follow the algorithmic procedures of such calculations we move beyond our intuitions and commonsense images.

To advance the illustration, we shall let the tree grow exponentially during its first four years, so that after one year it stands a foot tall, after two years 4 feet, three years 9 feet, and four years 16 feet, as in the following table:

$$x = (\text{time, in years}) = 1, 2, 3, 4$$
$$y = (\text{height, in feet}) = 1, 4, 9, 16.$$

If we want to discover the average growth rate, we simply divide the total height by the total years to obtain 4 feet per year. This would give us a straight-line graph, like the above case of $\Delta y = \frac{3}{4}\Delta x$. But the table reveals that during no single year did the tree grow at its average rate. In the first year, it grew only one foot; in the second year it grew an additional 3 feet; in the third year, 5 feet; and in the fourth, 7 feet, for its total of 16 feet. In fact the speed at which the tree grew changed constantly.

How, then, do we find out its rate of growth at any given moment? The brief, procedural answer: We repeat calculations of the average growth rate during smaller and smaller segments of time. This is the repetitive calculation that will eventually bring us into the terrain of the rhetorically unimaginable. To get there, we begin by mapping the information from the table into symbolic notation, and by establishing the functional relation between the symbols x and y. Because the tree's heights are the squares of the years it has grown, we express this relation with the equation $y = f(x) = x^2$, and plot it as a curved line (a parabola) on a coordinate grid, shown in figure 6.3.

Next, as with the previous example, we mark off a segment of the line in order to compare the intervals of change in height with those of change in time. We let P stand for the point of the growth rate we are seeking; its coordinates are (x,y) or (3,9) on the grid. Then we let Q stand for a second point. We obtain its coordinates by adding amounts indicated by Δx and Δy to the coordinates of P. This gives us $(x + \Delta x, y + \Delta y)$ as the coordinates of Q. Our original equation states that y is a particular function of x, that is $y = f(x) = x^2$. Our new pair of coordinates for Q, $(x + \Delta x, y + \Delta y)$, must likewise stand in the same functional relation to one another as the original equation. Therefore, $y + \Delta y = (x + \Delta x)^2$. This equation captures the relation between x and y, the years and height in the tree's growth, after intervals of change have been added to each of them.

At this juncture we perform some of Al-Khwarizmi's black-box, mechanical operations to obtain a new equation, $\Delta y/\Delta x = 2x + \Delta x$. (These and the following calculations appear in table 6.1.) We do this to transform the information of $y + \Delta y = (x + \Delta x)^2$ into a more usable, direct ratio of the changes in growth of feet (Δy) to those in passage of years (Δx). For readers oblivious to the rules of algebraic manipulation, the only point to bear in mind is that these are completely mechanistic operations and produce equivalencies at each step along the way.

Now we can see what happens when the change in the time segment (Δx) becomes smaller and smaller, which it must do in order for us to find the precise growth rate at the given moment, point P, the beginning of the fourth year.

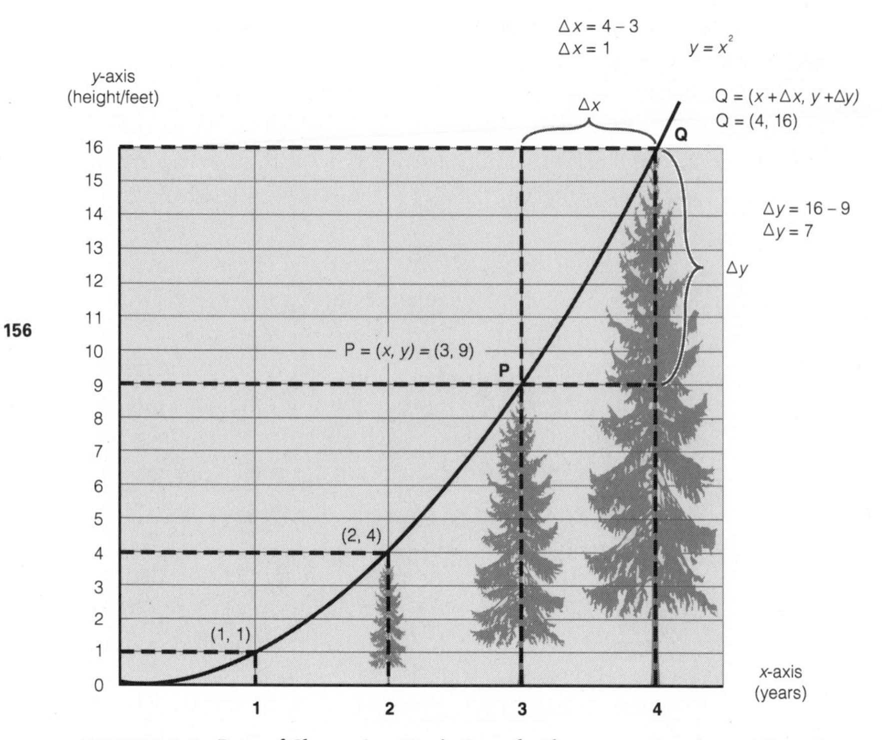

FIGURE 6.3. Rate of Change in a Tree's Growth, Shown as a Continuous Equation

Note that because we start counting from (0,0), the point (1,1) corresponds to the end of the first year and beginning of the second, much like figuring birthdays. Hence point *P*, (3,9), matches the end of the third year and start of the fourth. We know the tree has then reached an overall height of 9 feet, but how fast is it actually growing at that moment?

To find average growth rates for smaller and smaller intervals of time approaching point *P*, we first stipulate an arbitrary interval. Let Δx be 1, the span of one year, between years 3 and 4. Then Δy will be the change in height during the same span, or 16 feet – 9 feet = 7 feet. Our equation relates these values exactly: $\Delta y/\Delta x = 2x + \Delta x$; $7/1 = 2(3) + 1 = 7$. This gives us an average growth rate of 7 feet per year during the entire fourth year. But we seek the growth rate for only the beginning of the year, not for all of it. So we repeat the procedure, this time letting Δx stand for a smaller time period, say one-half (0.5) of the year. Now our growth in height will be $\Delta y/0.5 = 2(3) + 0.5$, or $\Delta y = 0.5\ [2(3) + 0.5] = 3.25$ feet per half year, which we double to obtain an annual rate of 6.5

feet. In other words, during the first half of the fourth year, the tree grows at an average rate of 6.5 feet per year.

If we stop calculating now, we would have found just the average growth rate for this half-year period, which would also mean a straight line for this portion of the graph. Yet we know our line is curved throughout because the tree actually grows at a constantly changing rate and never at an average one (or, more technically, does so for only one instant). So we must continue repeating our calculations of the changes in height during smaller and smaller intervals in order to approach more closely the actual growth rate at the beginning of the fourth year (see the calculations in table 6.1).

The concepts of mapping and function permit and explain each of the above individual calculations. But as we repeat the procedure of calculating these ever-smaller variables we are led to the third of the ideas uniting the operations of the calculus, the idea that there is a limit to a series of calculations. This is where we reach the doorway to the unimaginable. Notice, on the right side of the equation, as Δx approaches 0, the value of Δy approaches $2x$, and when $\Delta x = 0$, Δy is exactly $2x$. If, however, Δx actually reaches 0, then we will apparently have the taboo and impossible division by 0 because on the left side of the equation $\Delta y/\Delta x$ becomes $\Delta y/0$. It seems that *P*, the precise point we seek, which is called the instantaneous rate of change, cannot be. Thus did the calculus demonstrate for Voltaire "the art of exactly numbering and measuring that of which we cannot even conceive the existence."

It would appear we have arrived again at infinity's abyss, from which the Greeks had recoiled in horror. But Newton, Leibniz, and mathematicians ever since have simply found the limit or asymptote (convergence point) in calculating ratios of infinitesimally small variables, with 'infinitesimally' denoting quantities that approach, but never actually reach, zero. To conclude the example of our tree, as Δx approaches 0, Δy (the change rate in height) approaches 6 feet per year. This figure is the limit derived from repeating the calculating process, whence the technical term 'derivative'. The limit gives the conceptual endpoint or boundary beyond which a set of specific calculations cannot venture.

Leibniz named the procedure we have just followed 'differentiation', and he also introduced its more familiar notation, in which our example reads $dy/dx = 2x + dx$. The reason for the slight change in notation from the Δs to ds is subtle but momentous. The Δ symbol conventionally designates a specific, fixed difference between two quantities. The letter *d*, however, stands for the differential between two quantities: not a fixed difference, but the repetitive process of calculating smaller and smaller differences or infinitesimals.

TABLE 6.1

A Differential Equation

I. From figure 6.3:

$P = (x,y) = (3,9)$

$Q = (x + \Delta x, y + \Delta y) = (4,16)$

II. The steps in mapping the equation between *x* and *y:*

A. The general equation of the line, mapping any point *P:*	$y = x^2$
B. Mapping point *Q:*	$y + \Delta y = (x + \Delta x)^2$
C. Carrying out the multiplication of *B:*	$y + \Delta y = x^2 + 2x\Delta x + \Delta x^2$
D. Subtracting from *C* the original equation, $y = x^2$:	$\Delta y = 2x\Delta x + \Delta x^2$
E. Dividing both sides of *D* by Δx:	$\Delta y/\Delta x = 2x + \Delta x$

III. Differentiating—comparing changes in the growth rate of height (the *y*-axis) with smaller and smaller intervals of time (the *x*-axis):

A. With the time interval (Δx) set at one year (between years 3 and 4):

$\Delta y/\Delta x = 2x + \Delta x$

$\Delta y/1 = 2(3) + 1$

$\Delta y/1 = 7$

$\Delta y = 7'$ (the average growth for the entire fourth year)

B. With the interval of 0.5 year (between 3 and 3.5):

$\Delta y/\Delta x = 2x + \Delta x$

$\Delta y/0.5 = 2(3) + 0.5$

$\Delta y/0.5 = 6.5$

$\Delta y = 0.5(6.5) = 3.25$ (the average growth for 0.5 year, or an annual average rate of 6.5′ for this period)

C. With the interval of 0.1 year (between 3 and 3.1):

$\Delta y/\Delta x = 2x + \Delta x$

$\Delta y/0.1 = 2(3) + 0.1$

$\Delta y/0.1 = 6.1$

$\Delta y = 0.1(6.1) = 0.61'$ (the average growth for 0.1 a year, an annual average rate of 6.1′)

D. With the interval of 0.01 year, about 3.65 days:

$\Delta y/\Delta x = 2x + \Delta x$

$\Delta y/0.01 = 2(3) + 0.01$

$\Delta y/0.01 = 6.01$

$\Delta y = 0.01(6.01) = 0.0601'$ (the average growth for 0.01 year, an annual average rate of 6.01′)

We can continue adding 0s to the calculation ad infinitum, figuring 0.001, 0.0001, 0.00001, . . . units of a year. In so doing we will always approach an average rate of 6′ per year but never reach it exactly. This is the limit to our calculation, the tree's instantaneous rate of growth at point *P,* the end of the third year and beginning of the fourth. The differential equation $dy/dx = 2x$ results from differentiating $y = x^2$.

Technically, this procedure yields a first-order derivative or differential equation. It is often written slightly more elaborately as ($\lim \Delta x \rightarrow 0$) $dy/dx = 2x$ and says that as intervals of change in the independent variable *x* approach 0, the dependent variable *y* approaches 2*x.* Derivatives of greater orders can be taken by repeating the same process, packing more and more information into the equations. Newton used first-order derivatives to derive rates of velocity for moving bodies and second-order derivatives for rates of acceleration. With the notation evolved from Leibniz's practice, mathematicians usually write $f'(x)$, or $dy/dx = f(x)$, for derivatives of the first order and $f''(x)$, or $d^2y/dx^2 = f(x)$, for those of the second order.

This point cannot be overdrawn, for the richness and depth of the calculus, as well as of modern relational mathematics in general, issue from the fact that its symbols stand for procedures, not merely things or objects. It enables us to map ever-changing processes. The limit of the procedures in our example is $dy/dx = 2x + 0$, and so the equation becomes simply $dy/dx = 2x$. This is the derivative, or the differential equation, or the result from differentiating our initial equation of $y = x^2$. Decoding colloquially, it says the tree increases its growth in height twice as fast as it ages.

Whereas the notions of mapping and function stem from the cardinal principle of one-to-one correspondence, the idea of limit flows from the ordinal principle of recurrence. That principle says every number has a successor; there is no end to counting. By extension, it postulates the indefinite repetition of a procedure, once the procedure has been shown possible for a number and for its successor. The successor simply becomes the new number, and the procedure is repeated. If we can calculate the rate of change for a small interval, and repeat it for the next smaller one, we can continue ad infinitum, approaching a limit (be it 0, ∞—infinity—or a number), yet never reaching it.

With the calculus, therefore, the same principles of cardinal and ordinal number that had underscored the earliest expressions of numeracy become expanded and elaborated in the notions of mapping, function, and limit. The ability to calculate functions with limits marks the threshold of modern mathematical analysis. It signals, too, the parting of ways between our conception and our imagination, for although we can conceive the iteration of calculating techniques once their possibility has been demonstrated, we cannot often imagine the results in normal language. Powerful and accurate within its own symbolic domain, the "instantaneous rate of change" stupefies us outside it. A rhetorical oxymoron, the phrase makes no linguistic sense because 'change' requires time lapse, whereas 'instant' precludes it.

Indeed, both Leibniz and Newton had great difficulty imagining and defining just what an infinitesimal was, and at various points labeled it with all sorts of bizarre monikers, revealing its mystery, or at least its rhetorical orneriness: "ultimate ratio," "momentary increase," "fluent," "incipiency of magnitudes," "fluxion," "evanescent increment," "nascent augment," "amphibia between existence and nonexistence," or simply (this one from Leibniz) "a kind of fiction." Small wonder the divine Berkeley believed theology no less secure than the new math: "He who can digest a second or third fluxion . . . need not, methinks, be squeamish about any point in Divinity."

The advent of the calculus reveals a radical divergence of the symbolic language of analysis from the image-rooted rhetoric of classifying. This new, powerful, and abstract tool bifurcated the knowledge and wisdom that Descartes believed he had reintegrated through his *mathesis universalis.* Like Descartes, Leibniz too had professed his own grand and rationalist system, featuring God's preestablished harmony of substance and number. But for Leibniz, the fact that the techniques of the calculus could dispense with the necessity of imagination marked a great advance in human reasoning and learning.

Thus did Leibniz contribute ironically to the growing separation of analysis and taxonomy. Beginning with his youthful *Combinatory Art* (1666), he had even proposed among his earlier essays the *characteristica universalis,* a general, symbolic "language whose signs or characters would play the same role as the signs of arithmetic for numbers and those of algebra for quantities in general." By means of this universal language, he argued (to anyone who would listen), any information of whatever sort could be recorded systematically in abstract symbols and any problems in reasoning could be articulated and resolved. He called his formal method the "reasoning calculus" (*calculus ratiocinator*) and used the term "algorithm" to designate the general rules of manipulating abstract symbols. An indefatigable polymath, Leibniz also invented a mechanical calculator, which he believed would relieve humankind of much algorithmic tedium, freeing its reasoning capacities for more noble purposes (see plate 15).

By the mid–eighteenth century, the mathematical achievements of Newton, Leibniz, and a host of others had yielded the two critical components of the analytical vision and of the modern information age: formulas and algorithms. The first of these is evident in "classical" or Newtonian physics. Here the techniques of the calculus enabled natural and experimental philosophers, Newton chief among them, to generate new, abstract formulas that housed the invariants of knowledge. As philosopher Ernst Cassirer notes, the calculus became coextensive with science itself, and nature came to be viewed as governed and organized by a series of mathematically expressed relations, rarefied laws that described its mechanisms.

Newton authored the most famous exemplar, of course, with his law of gravity, $F = mm'/d^2$, or its equivalent, $F = ma$. This was an invariant relation between units of force, mass, and distance. Once formulated, it could be applied to any circumstance where relatively large bodies moved at relatively slow speeds. (Twentieth-century physics has added the boundary restrictions.) By means of such abstractions a mathematized physics epitomized the most exact and certain knowledge for the modern information era. Natural philosophers (eventually to be known as scientists) pushed science (still understood as exact, organized knowledge of natural things) across the "edge of objectivity" into the realm of symbolic abstraction, into the black box. Physics and astronomy had provided the beginning; they would be followed by other natural sciences. The most optimistic observers believed that the same methods would eventually embrace the moral and human sciences.

The second component of the analytical vision was yet more rarefied and had to do with the algorithms one followed. There was nothing fixed in algorithms save the rules of algebraic manipulation. Ever since the earliest reductions of Al-Khwarizmi these had been understood as purely "mechanical." Mapping, function, limit: The terms simply defined procedures that by themselves had no corresponding "reality" outside their own logic and rules. They were bound only by their internal coherence and by the virtuosity of the mathematical scientists, those performing artists who knew the music of number. The storage, management, and manipulation of information in abstract symbols (numbers, letters, points on a plane, or ultimately the "characters" of Leibniz) became thus a function of technique, of—we look far ahead—processing.

Yet in the eighteenth century, we must remember, analytic techniques remained subordinated to the service of representing the real, substantive world, the moral as well as natural sciences. They tendered the tools for the discovery and representation of the "there" out there. The very procedures of mapping, function, and limit were considered to be based on "self-evident" intuited truths about such realities out there as three-dimensional space, extended matter, causality, natural laws, and the like. Such intuitions enabled and emboldened d'Alembert and Diderot to imagine a world map, a coordinate system of information's geography and topography, whose realization lay at the heart of the *Encyclopedia.*

D'Alembert's Phantoms

In the Enlightenment's midlife Diderot and d'Alembert launched their Herculean plan: the editing and publication of what scholars now generally acclaim as the first modern encyclopedia. Eventually it reached seventeen huge folio (full sheet) tomes, appearing between 1751 and 1765, with an additional eleven volumes of plates following from 1762 to 1772. Other trees of knowledge had been devised earlier, such as that of Francis Bacon, and other encyclopedias had been published already, like that of Ephraim Chambers. But none carried such a massive volume of material or such a specific agenda. The *Encyclopedia* was to be the vehicle of progress. With it not only would the editors gather knowledge of various scientific and humanistic subjects spread throughout the world, but they would also, as Diderot announced, "explain the general system" of global knowledge for all "our nephews," his quaint term for the present and future generations of humankind.

Masters of all they surveyed, the editors surveyed all with a systematic spirit. They delivered to their subscribers and to the general public an unprecedented consolidation of knowledge, whose analytical arrangement, they believed, would resolve the problem of information overload. The alluring prospect of such resolution spurred them on, and into the task they poured the energy of youth and the righteous confidence of the Enlightenment. Their labors would cease only with the complete recasting of knowledge.

Two themes set in high relief how Diderot and d'Alembert went about this enterprise. First, through a series of articles and commentaries (part of the *Encyclopedia* itself), the editors explicitly spelled out the analytical vision of knowledge for their contemporaries. This was largely the work of d'Alembert. Not only did he rank as one of the age's premier mathematicians, but he also summarized effectively the process whereby information had emerged from the symbolized language of modern numeracy. In so doing he depicted both the abstraction and the reductionism of the new information technology and idiom.

Second, the editors together applied the analytical vision in their creation of a new reductionist hierarchy of organized knowledge, an "encyclopedic arrangement" that underlay all the information contained in the *Encyclopedia.* For the sake of their readers' convenience the editors opted to present topics in the volumes alphabetically, in dictionary fashion. But although the dictionary might help one look up a subject, one needed to see its overall place in the world map of cognitions in order to comprehend it. The reductionism inherent in modern numeracy made this possible. The analytical vision, therefore, would come to offer a new definition of knowledge and to provide a new means of closure to the exhaustive process of gathering information.

The editors began with the premise that the general system of knowledge had to be securely founded on "first principles, general notions, given axioms." These were the "roots" of the tree of knowledge and the key to its organization. Also termed "metaphysics," the most "primitive" (foundational) of the first principles were the two concepts grounding analysis itself: abstract number and arithmetic function. Both these concepts lay behind the techniques and procedures of the calculus and were taken directly from the cardinal and ordinal characteristics of number, correspondence and recurrence. Through abstract number and arithmetic function other features of analysis and, subsequently, of encyclopedic order would be promulgated.

In numerous passages d'Alembert reflected on the abstract nature of the calculus to provide an account of analysis. Rhetorical descriptions had long depicted analysis as the "resolution" of "a whole into its parts," a meaning evolved from the Greek *analyein* ("to break up"). But with his often reiterated, mathematical definition, d'Alembert recast the conventional treatment. Strictly speaking, analysis was the "method of solving mathematical problems by reducing them to equations." It was thus identical to algebra, as Viète had first suggested. By d'Alembert's day analysis and algebra were used interchangeably, and either one designated the discipline Newton had called "universal arithmetic." Moreover, both terms included the calculus as the general means of calculating quantities or magnitudes (also synonymous terms in the eighteenth century), and of representing them by abstract signs, usually letters of the alphabet.

First glimpsed by Descartes, the novel concept of abstract number, which lay at the core of analysis, had been defined much more precisely by Newton as "an abstract relation of one quantity to another," rather than as a multitude of units. D'Alembert showed how the abstract nature of these relations carried one far beyond the sense-laden imagery embodied by words and classifying, which for him was only a halfway step to the more thorough-going abstraction intrinsic to modern numeracy.

Abstraction designates a mental process or "operation," d'Alembert stressed, not a fixed category or thing or object. Like the ancients we too begin with things and our sensations of them. But they had proceeded from things to sensations to essences, or "substantial forms," d'Alembert noted. In this process they considered sensory images to be the informational content expressed and classified in the nouns and adjectives of language. We moderns, by contrast, pay no attention whatsoever to the sensory images that make up the presumed essences of objects. Instead, we "divest matter of almost all its sensible properties" and consider only a very "particular property" in any given object, envisaging in it "only its phantom."

Abstract containers of information, these phantoms are lodged in the symbols of analysis—a point, a number, a letter, an operational sign. Just as "there is no [intrinsic] connection between each sensation and the object that occasions it," so too is there no internal connection between these phantom-symbols and matter. "Only a kind of instinct" compels us to "leap the gap," from objects to sensations to phantoms. Here the process of abstraction is purely mathematical, and d'Alembert's rhetoric captured what we have already identified as the cardinal feature of abstract number. An extrinsic, cardinal correlation, mathematical mapping joined the phantom of the object and the material world, uniting information and reality in symbols.

But the phantoms offered more than discrete, abstract containers of information. They were also abstract relations, and as such they entailed magnitude and the idea of counting or serial order. This is the ordinal principle of recurrence, which allows us to pass from one number to its successor. "Thus," d'Alembert summarized pithily, "the number 3 expresses the relation of one magnitude to another, smaller one that is taken as *l'unité,* and which the larger contains three times. On the contrary the fraction 1/3 expresses the relation of a certain magnitude to one larger that is taken as *l'unité,* & which is contained three times in this larger magnitude." In this passage *l'unité* reveals the idea of ordinal number, for it refers to the total or complete unity of mathematics, which is nowadays termed "density"—that, as we noted in the last chapter, between any two numbers on a continuum one can always find a third. In contrast to English, which distinguishes the discrete 'unit' from the dense 'unity' in mathematical relations, the French *unité* has historically embodied both discreteness and density. It thus served d'Alembert as a shorthand reference to the entire phantom world of abstract relations, the world of number.

D'Alembert grew even more expansive. Because numbers are abstract relations perceived by the mind and distinguished by particular signs, the "science of numbers" reigns as the art of combining relations between numbers, or to add an important convolution, relations between relations. Finally, in its most general meaning, quantity—the entire number realm—concerns "everything . . . susceptible to augmentation or diminution." Thus did his reflections upon the concepts and techniques of the calculus lead d'Alembert to postulate and articulate a purely abstract realm of phantoms, symbols, and their operations, what we have been calling the analytical vision. This phantom universe comprised the abstract formulas that mapped or stored the invariants of knowledge. It framed and mediated the contents of our exchanges with the outside world.

Abstract number provided the foundational first principle of analysis. The second cornerstone was arithmetic function. It followed directly from the idea of abstract number and allowed one to devise "certain rules relative to the operations" that "one could perform on [abstract] numbers." These rules were the "rules of arithmetic," by whose governance one manipulated abstract relations. They too could be effectively reduced because all mathematical operations followed from the simple plus and minus: Multiplying and dividing, the most obvious examples, only provided "abbreviated manners" of adding or subtracting. The same held true for the more complicated procedures, from roots and exponents to trigonometric functions to the ratios of differential calculus.

In anchoring reductions, the two operational rules of addition and subtraction were entirely general and in fact could be performed within all "particular arithmetics," whether a decimal system, Leibniz's binary system, or any other system of number signs. The rules comprised "an art of making combinations without knowing the numbers one seeks." They thus signified a way of expressing "numbers by the different characters of numerical characters." With this rather cumbersome phrase, d'Alembert voiced a simple but profound idea. The rules of arithmetic govern the manipulation of letters or any other symbols, and in turn symbols correspond to any number signs. Otherwise stated, the simple rules of arithmetic generated the algorithms needed to manipulate the phantoms.

D'Alembert went no further in justifying analysis. Nor did he feel the need to, for these foundations seemed so self-evident. They comprised what, as we noted above, were the two critical components of the modern information era: formulas and algorithms. From these bases we are led to construct the entirety of pure mathematics and to posit its connection to the material world. From there we proceed to the rest of the sciences, as the dense, phantom universe with its unending permutations gives our experience malleable shape.

Both d'Alembert and Diderot believed that the foundational concepts of abstract number and arithmetic function were axioms. Axioms were the necessary first principles of arithmetic, those "general properties of relations" existing independently of the particular "signs" by which one expresses specific numbers or specific relations. More important, axioms made it possible to proceed from pure mathematics to the other sciences and branches of organized knowledge. They made it possible to generalize, and to do so in a new analytical manner that diverged markedly from generalizing as traditionally linked with the hierarchies of classification.

D'Alembert's treatment of the term "general" itself conveyed most clearly the new process of generalizing. Traditionally, he wrote, the adjective 'general' had typically meant something common to everything considered under the "same point of view." Thus in physics one said that weight is the "general" property of matter and in metaphysics that sensibility is the "general" property of animals. In these examples 'general' refers to a common quality or property possessed by all the members of a given class of objects. Etymologically, the French *general* and *genre* ("genus") both derived from their common Latin forbear *genus* ("birth, race, kind"); *general* was merely the adjectival manner of referring to any substantive class or genus to which objects belonged.

When it came to depicting mathematical generalization, however, d'Alembert proposed a radically different meaning for the term. One calls a theorem or a problem "general" when "a large number of consequences and applications" result from it, or when a theorem spreads out almost to an entire science. Generalizing here meant applying a formula, which includes "an infinity of cases" and from which "one can derive many other particular formulas." With a formula—a "general result drawn from an algebraic calculation"—one need only substitute particular numbers for letters in order to find results in any appropriate case, an "easy method" to use, d'Alembert reassured skeptical readers. Finally, if a formula can be rendered "absolutely" general, it offers the greatest intellectual advantage, often enabling one "to reduce an entire science to a single line" of symbols.

Mathematical generalization, in brief, meant mapping to a domain of specific cases the abstract, functional relations found in a formula. Though designated by the same word, 'general', this conception differed dramatically from the substantive features or sensible properties that applied to each member of a genus. General properties here were operational, and the axioms on which mathematical generalization rested were simply its fundamental operations. Abstract number and function, the most general of all axiomatic, first principles, "spread" their operations throughout the entire domain of mathematics. Thus did the axioms of analysis provide d'Alembert and Diderot the clearest example of how proper reasoning proceeded from first principles. Thus, too, did both the image and technique of operational reduction underlie organized information. These two beliefs allowed them to envision the world map as a mathematized hierarchy of knowledge.

The Encyclopedic Arrangement

For both editors the same first principles or axioms that grounded analysis also served as a foundation for the system of organized knowledge. They believed these principles issued from the faculties of the human soul or mind. In this assumption they took their initial example of organized knowledge from Francis Bacon, who had sketched a tree of learning along the lines of the soul's three major functions: memory, reason, and imagination. Similarly, Diderot and d'Alembert held that these three functions begat, respectively, the three major divisions of knowledge—History, Philosophy, and Poetry (or the arts in general). At this juncture, though, similarities to their empiricist forefather ended; the divide of mathematics separated them from Bacon and his age.

Bacon's consideration of the soul's parts had derived from his own classificatory vision of natural order. A century and a half later d'Alembert considered

TABLE 6.2
The Mathematical Topics of the Encyclopedia as Listed in the "Detailed System of Human Knowledge"

MATHEMATICS		
I. Pure Mathematics	II. Mixed Mathematics	III. Physico-Mathematics
A. Arithmetic	A. Mechanics	
1. Numerical arithmetic	1. Statics	
2. Algebra	a. Statics proper	
a. Elementary	b. Hydrostatics	
b. Infinitesimal	2. Dynamics	
i. Differential [Calculus]	a. Dynamics proper	
ii. Integral [Calculus]	b. Ballistics	
B. Geometry	c. Hydrodynamics	
1. Elementary (military architecture, tactics)	i. Hydraulics	
2. Transcendent (theory of curves)	ii. Navigation, naval architecture	
	B. Geometric astronomy	
	1. Cosmography	
	a. Uranography	
	b. Geography	
	c. Hydrography	
	2. Chronology	
	3. Gnominics (art of constructing dials)	
	C. Optics	
	1. Optics, proper	
	2. Dioptrics (perspective)	
	3. Catoptrics (properties of mirrors)	
	D. Acoustics	
	E. Pneumatics	
	F. Art of conjecturing (games of chance)	

the soul's "metaphysics" as purely mathematical. He was explicit and insistent. All three branches of human understanding, he urged, could be reduced to a single set of analytical principles, based on abstract number signs and the rules of arithmetic governing their operations, the very same principles that had tendered the foundations of analysis. This was not merely trimming the Baconian tree of knowledge, as one historian has written; this was replanting it in mathematical soil. Leaving the metaphor aside, it meant a radically new basis for organized knowledge.

A large puzzle in the "Detailed System of Human Knowledge" allows us to gain entry into just how the editors pictured this new hierarchy of organized knowledge. The puzzle rises from the placement of the calculus in the general category "Mathematics." We see on the chart (table 6.2) that both the integral and differential branches of the calculus fall under an apparent hierarchy of increasingly general headings: "Infinitesimal," "Algebra," "Arithmetic," and "Pure Mathematics." At first look, this hierarchy appears to be a rather traditional sort of taxonomical arrangement.

Yet why would the calculus be identified as a specific subcategory of a more general arithmetic or algebra? After all, arithmetic is not more "general" than the calculus; to the contrary, generality flows the other direction, as it were. Although an arithmetic equation ($2 + 3 = 5$) remains restricted to the realm of numbers, an algebraic formula ($a + b = c$) serves to express an infinity of arithmetic calculations. The calculus embodies an even greater collection of mathematical problems, operations, and applications than any of the categories in which it resides: A single line of calculus, d'Alembert reminded his audience, allows one "to learn entire sciences in a short time."

This curiosity can be explained only by realizing that the links between mathematical topics were operational, not generic, despite the topics' superficial taxonomical display. Arranging topics has now shifted away from a classifying hierarchy to one of mathematical reduction. Placing a mathematical subject under a heading meant reducing that subject to the more primitive operational principles of the heading, rather than subsuming it under a general category in the manner of genus and species. (Similarly, recall that Viète had superimposed "scalar magnitudes" over the separate genera of arithmetic and geometry.)

Both differential and integral calculus were listed under "infinitesimal" because both utilized calculations involving an "infinitely small quantity." In turn, infinitesimal calculations reduced to or presupposed the operations of algebra, the "method of calculating indeterminate quantities," designated by letters of the alphabet. Those functions reduced further to the properties and functions of "abstract quantity," and hence to the foundational, abstract relations of arithmetic and "pure" mathematics. At each step in this reductive schema, from integral calculus to pure mathematics, the inference was derived instrumentally from the logic of mathematical operations, not from taxonomical classes. And "the more one reduces the number of principles of a science, the more one gives them scope." Such a reduction increases generality, understood as the ever-broadening application of mathematical functions; it constitutes the true systematic spirit.

From pure mathematics the editors extended this vision of mathematical reductionism to a category they called "mixed mathematics," which introduced "facts" or information from the world. Mixed mathematics systematically employed mathematical abstractions and remained tied axiomatically to the first principles of pure mathematics. Nonetheless it interspersed observations, even "occult powers" like action at a distance, throughout its claims. Newton's law of gravity, for example, was for d'Alembert an experimental truth described by mathematical formulas, rather than a truth that could be rationally demonstrated from purely axiomatic, mathematical principles. It followed that mechanics and with it much of classical, Newtonian physics were subsumed under "mixed mathematics."

If ever Newton's laws could be derived purely from axioms, d'Alembert believed, then they would constitute a "rational mechanics." Such a mechanics would be a purely mathematical reconstruction of the laws of physics, deriving strictly and logically from axioms. In this event the results of observation would agree entirely with a priori reasoning, and the laws of physics then would be necessary, not contingent. A completely rational mechanics would be identical in logical rigor to pure mathematics, which d'Alembert also termed the study of "general physics."

The subjects of "mixed mathematics" thus eluded complete, rational reconstruction. At a future date, when all the facts were in, the problematic portions of these subjects would be reduced to a set of necessary laws. Until such time their axiomatizations would remain correct as far as they went, but incomplete. Separating pure and mixed mathematics then lay with the possibility of constructing mathematical disciplines solely from the general concept of mathematized matter—the phantom world—itself derived from, but unalloyed with, any sensory perceptions. Mix in the senses, mix the mathematics.

Beneath mixed mathematics the editors added another level of organized natural sciences, which they termed "physico-mathematics." Into this category fell the sciences that had not yet been rationally reconstructed to the extent of classical physics. Still, these sciences utilized some mathematical calculations in their experiments, and therefore some of the operational principles of analysis. In actual practice the distinction between mixed and physico-mathematics was quite murky (as is suggested by the empty column under "physico-mathematics" on the "Detailed System"). Even though they essayed at times to separate them, the editors probably intended the categories as roughly interchangeable, since the placement of any science in a category depended upon the degree to which it had been mathematized. This involved a particular judgment about the current status of a subject, which could change quickly.

It is hard to overemphasize the importance of the reductionism of analysis for the editors as they considered the organization of knowledge. Once in place, the links of organized knowledge reached to the nonmathematical, "particular sciences" and from there to the practical and liberal arts, and even to the moral sciences. The details of this reduction grow murkier step by step, as one proceeds from "General and Experimental Physics" (also at times termed "particular physics") to "experimental philosophy," a far vaguer category that included such topics as morality, public law, and history, subjects more traditionally aligned with the moral sciences.

Behind the increased vagueness at each step downward in the reductive hierarchy were phenomena whose causes still lay outside reason, and whose chains were not perceived, or were perceived only "very imperfectly." Such, for example, were the phenomena of chemistry, those of electricity, those of the magnet, and a host of others still beyond the reach of analysis. These were subjects where the scientist initially ought to seek out the facts. The more he accumulated them, the closer he would be to seeing their connections to one another and to the existing body of axiomatized knowledge, and the closer he would be to giving mathematical form—or formula—to information from the world. The same held true for the moral sciences, all the subjects "that belong in a certain sense to experimental philosophy, taken in its extension."

With its reductive hierarchy of mathematical operations, the analytical vision of knowledge structured the entire encyclopedic arrangement. Science as organized knowledge (*scientia*) was now derived from "science" as operational method, the algorithmic manipulation of symbols that mapped reality. Diderot summarized: "If the perhaps infinite sum of the possibly infinite multitude of nature's molecules were completely known to us, we should see all phenomena occurring according to rigorously geometrical laws." A mathematized taxonomy meant that all the differences between topics were marked by degrees of mathematical reduction, not classificatory kinds. The farther one was removed from the operational foundations of pure mathematics, the more the categories of knowledge varied. They became looser, less assured, and increasingly arbitrary, for fewer links could be established between existing information and axioms. Much more remained to be discovered. Conversely, the more any given discipline became rationally reconstructed with chains that tied it to first principles, the more its place in the encyclopedic arrangement became mathematically defensible and secure.

Analysis Triumphant

At the beginning of his article on the elements of science, d'Alembert had summarized the analytical vision that underlay all possible knowledge. We begin with the propositions that form any science, he stated, and we consider these propositions in the "most natural and rigorous order possible," supposing that they formed an "absolutely continuous series," with each proposition depending uniquely and immediately on its antecedents. Everything in a single science thus can be reduced to a "first proposition," which we call the "element" of the science in question. In turn, he added, "the elements of all the sciences are reduced to a unique principle, whose chief consequences would be the elements of each particular science."

In this fashion the human mind would come to see all knowledge as unified by a singular perspective: "The universe, if we may be permitted to say so, would only be one fact and one great truth for whoever knew how to embrace it from a single point of view." The only difference between humans and God is that God would perceive all the objects of knowledge at a glance, whereas humankind would need to reason them out. As humans we cannot see the entire chain of knowledge; we do not even see all the links constituting each science.

Yet, the reductionist hierarchy made possible by the analytical vision meant that, also as humans, we did not need to see each link immediately. Reductionism provided closure to the process of exhaustive information gathering, in the same way that taxonomical hierarchy had precluded the need to name all the things of the world. Analysis would allow us to distinguish which general truths or propositions serve as the base of others, to know what a piece of knowledge looks like when we have got it, and to see where it fits in the overall structure of knowledge. Indeed, the information of the world becomes knowledge once it is reduced and placed into the overall analytical structure and organization. Future generations of "our nephews" could continue to fill lacunae as they saw fit, secure the links within and between the various disciplines, and move toward the ultimate cognitive goal: the analytical world map.

By the end of the eighteenth century, the modern information age was in place. Its central elements consisted of the book and the printing press as the chief technology of communications. Initially this technology had provided the catalyst for an information explosion that accompanied the Renaissance recovery of antiquity, the New World discoveries, and the Scientific Revolution. The volume of information buried in millions of new books had inundated traditional classification schemas and even seemed to render the idea of classifying completely unworkable. There was no "bean in the cake," Montaigne had said. No empyrean heights permitted squalid little humans to survey and order with any assurance such immense fields of knowledge.

Humankind now needed an innovative, information technology, a compass. Francis Bacon understood this need for navigation. The real compass, however (of which he had little grasp), lay with the new symbolic language of analysis, first devised in mathematics and then extended to other reaches of thought. This was the language that ordered knowledge. By the eighteenth century the analytical vision had anchored a new way of thinking, using the symbols of the calculus as its primary and most powerful instrument. With this language classical physics had been formulated, and with the *Encyclopedia* analysis itself was seen as the source of information discovery, storage, and organization.

Like other information ages, the modern age issued its own ironies amid its own achievements. Diderot and d'Alembert both thought that by extending the systematic spirit one could know what there was to be known about moral and human matters as well as about nature. On the face of it there was no reason why knowledge in the moral sciences should differ in kind from that of the natural sciences. Except for the niggling bugbear, Diderot came to see, of free will. Mathematical science could inform phenomena of the world with its determined mechanisms, but linear predictions from formula did not admit the intervention of whimsy in the logic of their calculations. And without whimsy people could not be moral or human. Knowledge required determinism; action entailed freedom. At the very moment of the analytic vision's triumph, this core dilemma gave birth to the Romanticist reaction, and with it the separation of the two cultures, scientific and humanistic.

These ironies notwithstanding, the modern information age would continue, just as classifying has done, even into our own times. Indeed, the aging of information in our narrative does not represent a linear, chronological sequence, whereby one age follows and replaces another. Rather, it bespeaks the increasing ability of humans to abstract themselves from the immediacy of their surroundings and to devise different and distinctive manners of structuring their experience. The classificatory potential in alphabetic literacy culminated in creating the first such moment in the classical information age. Its acme, Aristotelian *sophia,* stood atop taxonomical *epistēmē,* yielding a vision of "science" as organized knowledge. It was the age of wisdom. With the *Encyclopedia,* the modern age reached its zenith, realizing the analytical possibilities of a new "science," whose methods were rooted in numeracy. Afterward, of course, huge amounts of information crossed the frontier into its reductive, encyclopedic realm, and still do. Yet its vision of information has stayed the same. It is the age of knowledge.

PART III

The Contemporary Age of Computers

Analysis Uprooted

Words are signs. . . . A sign is an arbitrary mark.
—GEORGE BOOLE, *An Investigation of the Laws of Thought*

Babbage and the Analytical Engine

If the dream of René Descartes's ushered in the modern information age, the vision of the affable British eccentric Charles ("Sir Alphabet Function") Babbage (1791–1871) presaged our contemporary information age. Mathematician and engineer, political economist and reformer, "operations analyst" and salon host (party-goers baptized evenings at his home "doing our Babbage"), this quirky inventor envisioned among his many endeavors the first modern computer. His life's most passionate project, he named it the "analytical engine" and conceived it initially as a device to deal mechanically with the monotony of astronomical computations and calculations.

Far surpassing earlier attempts at mechanical calculators, such as Pascal's adding machine or even Leibniz's much more intricate multiplier, Babbage's device consisted of two parts: "1st. the store in which all the variables to be operated upon, as well as those quantities which have arisen from the result of other operations, are placed. 2nd. The mill into which the quantities about to be operated upon are always brought." In contemporary parlance, the store was to be the storage (memory) of internal information (which he referred to as "numerical data"); the mill, or central processing unit (CPU), would perform the operations or functions to which the information would be subjected (see plates 16 and 17). He also devised a punch-card system, based on Jacquard's looms, for entering information and giving the engine its operating instructions.

In drawing the plans for his machine Babbage simply made explicit the two functions of analysis that had accompanied the rise of symbolic mathematics during the modern information age: the storage of information in abstract symbols (numbers, letters, or points on a plane), and the algorithmic operations by which the symbolized information was manipulated. As noted before, ever since the techniques of Al-Khwarizmi had been introduced to quantitative thinking, algebraic procedures themselves had been understood as largely "mechanical" (their invention and application far less so, more a function of intuition or insight or imagination). Now in Babbage's workshop, staffed with the finest machinists and draftsmen of the 1830s and '40s in England, the mechanism of the intellectual techniques associated with the abstract language of number would become a physical reality as well . . . if only funding could be found.

But alas! Technically feasible, the analytical engine remained economically and mechanically dubious and was never built, despite Babbage's perennial search for investors and even though working portions of an earlier model, the "Difference Engine No. 1," had already been constructed. Nevertheless, that Babbage could even imagine in such detail a machine to do the mechanical, algorithmic work of analysis and information storage testifies to the analytical vision's ascendancy and mastery over the scientific and technical world. From the *Encyclopedia* forward, the global project of mapping the informational universe extended its conquests by means of greater and greater abstract leaps. Conversely too, abstraction itself came to mean predominantly the subsumption of information into manipulable symbols.

In extending its reach, the analytical vision soared into the heady, dizzying heights of mathematical physics, stretching thinly its connections to the sensible world of space, quantity, and matter. During the eighteenth century, the calculus had mapped information about nature into a symbolic language, one that transcended words and things. But even so, the world had always been intuitively understood. It was out there, still within the realm of common sense: space had three dimensions; quantity made up natural number; cause and effect could be experienced; matter could be touched, handled. The techniques of analysis mapped the information rooted in this palpable world.

By contrast, the collective genius of nineteenth-century physicists and mathematicians created abstractions of mind-boggling sophistication that drew out the connections to the sensible world so far that only a few specialists could grasp them. Mere mention of the new, non-Euclidean geometries of Georg Riemann (1826–66) and others illustrates the point. These geometries challenged the centuries-old, self-evident belief that the universe was composed of a real, three-dimensional space in which some lines were straight and parallel. Riemann even postulated as an axiom that lines could never be straight or parallel. He then explicated the subsequent lemmas and procedures for manipulating an everywhere-curved space. Initially a mathematician's mercurial delight, this logically consistent, though scarcely imaginable, curved space eventually found a home in Einstein's general theory of relativity.

The creation of non-Euclidian geometries typifies the growing abstraction and formalism of mathematical thinking in the nineteenth century. In a long and steady process of detachment, mathematics became uprooted from its earlier, commonsense and intuitive foundations. Step by analytical step, it pared away its connections to the world and increasingly sought its vocation only in devising explicit postulates and rules of procedure, even though these might be quite arbitrary. (The axiomatic "law" of noncontradiction nobly withstood the trend.) Mathematics became the "science of order" (sometimes now titled a "postulational-deductive science"). It analyzed implications, rather than demonstrated global truths from self-evident axioms.

Where mathematicians went, physicists often followed closely in tow. Sometimes they even led the way. With abstract postulates they too created their own worlds: infinitely silent, conceptual universes comprised of bizarre entities such as mathematical "particles" and of time-warped areas such as "phase space." Unlike the abstractions of mathematicians, the physicists' postulated systems retained an all important connection to our haptic, phenomenal world by means of experimental testing of predictable events. Nevertheless, the subtleties of their formulas far transcended ordinary, visual imagination, as physicists mapped more and increasingly abstract information into their symbols and operations.

With mathematics and physics, then, analytical abstractions grew ever more attenuated throughout the nineteenth century, both from the tangible world and from natural language. This rarefaction stripped away many of the assumptions that previously had linked d'Alembert's phantom world of analysis to outside realities. In turn, as abstractions became more tenuous, as well as attenuated, the whole question of their connectedness to the world began to worry some investigators. This led a few to focus their attention more directly on the logical "chains of reason" that held abstractions earthbound and together. Such studies initially promised a new foundation for analysis, derived solely from the postulates and assumptions of a newly symbolized logic. Not only would symbolic logic ground more securely the abstractions of scientists, but of equal importance, it would connect their universe to the words and things of ordinary language.

Two logicians in particular, Augustus De Morgan (1806–71) and George Boole (1815–64), pursued this line of inquiry. They sought to algebraicize both logic and natural language and hence accomplish for language what the creators of modern mathematics earlier had achieved for numeracy: establish it on a completely symbolic basis. These investigations would uncover the ultimate foundations of knowledge, Boole wrote, in the "laws of the symbol," which were the "essential laws of human language," and in their underlying "operations of the human mind."

Descartes and Leibniz had only dreamed of such a symbolized natural language with their talk of "primitive notions" and *characteristica universalis.* Diderot, too, had fantasized about a symbolic, "reasoned alphabet." De Morgan and Boole leant reality to the dream. For those who followed (giants like Frege, Cantor, Dedekind, Peirce, Peano, Russell, Whitehead, and others), their work provided a focal point of prodigious efforts directed toward expanding the reach of symbolic logic and establishing mathematics on what were believed to be solid, logical bases. Held together by the assumptions and rules of formal reasoning, this new logical reductionism would reconnect analysis to the world it purported to map.

With mathematical and logical analysis joining forces in the second half of the century, confidence in the powers of symbolic reason stood like the mythical Atlas, shouldering the heavenly, abstract vault for scientists, logicians, philosophers, and other intellectuals. Yet in the end, this confidence proved to be a Victorian conceit. For the efforts to ground the analytical vision in symbolic logic did not secure new foundations for its techniques. Nor did these efforts yield a new, philosophically assured means of linking analysis with our ordinary experience. Instead, the techniques of analysis themselves became severed from their foundations—uprooting formal symbolic reasoning at the very moment of its creation.

Ironically, Boole's brilliant technical innovations prepared the way for this uprooting. But not until the dawning of the twentieth century was the irony exposed, and then quite unexpectedly from two sources. From outside the rarefied world of logic, the empirical investigations of physicists uncovered the relativity of space-time and the indeterminacy of the quanta. These discoveries radically altered the classical picture of a deterministic, law-driven universe.

Even more dramatically, within the world of analysis the abstractions of a mathematized logic began to turn inwardly on themselves, imploding the world map of organized knowledge and revealing contradictions and limits to the claims promulgated by analytical techniques. These developments would culminate in the class paradox of Bertrand Russell and the incompleteness theorem of Kurt Gödel, two discoveries that ripped out the final conceptual roots of analysis. After them the confident belief in philosophical "foundationalism" of any sort faded as the techniques of logical, mathematical, and linguistic analysis undermined all attempts to ground knowledge in the global, self-evident truths so dearly beloved by the eighteenth century.

For the remainder of this chapter, we shall explore the growing attenuation of the analytical vision, Boole's ironical attempt to ground it in symbolic logic, and the final uprooting of analysis. This will serve as a prelude for appreciating more immediately our own contemporary information age, so centrally allied with computer technology. Our age inherited the visions of the encyclopedists with their world map and of Babbage with his plans for an analytical engine. We inherited as well countless techniques of abstraction and symbol manipulation as the instrumental means for converting these visions into material realities and for mapping the real world of things and their motions. But even as the techniques were being created, their connection to the outside world stretched beyond the breaking point. For us, the presumed content of the world map no longer carries conviction. With the uprooting of analysis, the world map—the chart of reality itself—has been transformed into the world of pure technique: knowledge transformed into information management.

The Attenuations of Mathematical Physics

In the mid-eighteenth century, d'Alembert had centered the analytical vision on the "method" of solving problems by "reducing them to equations." Further reductions would lead one to the fundamental "elements" of any given science and ultimately to the "unique principle" of all the sciences, the "single point of view" by which the entire universe could be seen godlike as a unitary "fact." The discovery of knowledge was thus reduced to the algorithmic procedures of analysis, while the organization of knowledge was tied to invariant, substantive relations, those laws of nature discovered by the procedures. In the following century, mathematical physicists continued to expand greatly the technical proficiency and reach of their analytical tools, and along with them confidence in the abstract, reductionist, and determinist features of the analytical vision.

This was the great age of "scientific determinism," the doctrine that the present position and momentum of any given particle, or system of particles, completely and unambiguously determine all future positions and momenta of the same particle or system. If one knew the positions and momenta of all the particles in the universe, it followed, then the entire future could be predicted for all time. Later in the century, the brilliant French mathematician Henri Poincaré (1854–1912) would summarize scientific determinism trenchantly. The connection between the present and future is "law," he wrote, and law is "a differential equation."

Increasingly identified with the techniques of calculus, analysis not only permitted deterministic predictions of myriad natural phenomena, it also enabled scientists to compress more and more information into its abstract symbols. With both the cardinal and ordinal principles of correspondence and recurrence connecting symbols to experience—at key points, minimally—and generating ever more sophisticated algorithms, physicists and mathematicians could map information into abstractions at greater and greater removes from perceivable, physical reality. These abstractions lay totally outside the pale of human imagination and natural language. Mathematical symbols not only acquired a life of their own, they also became the specialized, formal language of the new information technicians.

The pinnacle of these developments can be found in the work of Ireland's "greatest scientist," Sir William Rowan Hamilton (1805–65), whose accomplishments will illustrate for us the extraordinary abstractions of mathematical physics. His work extended the achievements of two other premier mathematicians of this "golden age" of physics, Joseph Louis Lagrange (1736–1813) and Pierre-Simon Laplace (1749–1827), whom we met earlier. All three men broadened and deepened the procedures and findings of classical physics, providing rigorous, analytical expression for the Newtonian world map. For them, differential equations, the keys to scientific determinism according to Poincaré, comprised the language of analysis in physics. In fact, many new techniques of calculus had been developed directly in response to problems of physics and had become far more complex by Hamilton's time.

Chief among them stood the ability to manipulate equations with three or more variables or dimensions. During the mid-eighteenth century, d'Alembert, Euler, the young Lagrange, and several others had begun grappling with such calculations when their attention was piqued by "three-body" problems in astronomy and mechanics. How, for example, could one mathematically account for a satellite like our moon, which besides its own inertial force is tugged at gravitationally by both the earth and the sun?

Some segments of these problems could be solved with what later came to be called "ordinary" differential equations. But Lagrange systematically developed new techniques of "partial" differentiation, which wove together the strands of many contemporary investigations and provided general methods for solving problems of three and more variables. His accomplishment marked a stunning achievement in formulating the abstractions of mathematical physics on which Hamilton and others after him based their discoveries. The procedures of partial differentiation still stand today as some of the most powerful analytical tools used by scientists to elicit and control specific information about natural processes. As such, they here require some further explication.

We have already seen how the calculus captures information about fluctuating or varying rates of change. A differential equation takes a fixed, invariant relation, such as a trigonometric function or an algebraic equation ($y = x^2$), and sets it in motion, so to speak, with calculating procedures ($dy/dx = 2x$). In our elementary example from the last chapter, we plotted this process on a coordinate system in order to comprehend the relation between changing rates of growth in a tree's height and age at any given instant. In so doing we abstracted the tree's three-dimensional, leaf and bark, physical features for the purpose of mapping its growth on a coordinate system of two dimensions, as though plotting it in flatland. This procedure works for some kinds of mappings, but capturing the phenomenal flux in our normal three-dimensional world of space more often requires minimally a third coordinate or axis.

Typically this coordinate is designated as z and is conceived as passing at right angles through the plane of the x- and y-axes at their point of origin or intersection (0,0). Now, instead of imagining a coordinate system as a two-dimensional grid or street map, we must envision it as a cube with equally spaced, parallel grid lines extending from an origin point (0,0,0) in three, right-angled directions, as shown in figure 7.1. Three coordinates (x,y,z) will be needed to determine any single point within the cube, rather than the two used previously in the plane coordinate grid. Further, plotting a particle's motion in three-dimensional space, or mapping flow changes simultaneously in three dimensions (as with our tree—it ages, it grows taller, it expands its girth) requires equations of three variables, or even more. Changes in one dependent variable then become a function of changes in two or more independent variables.

To calculate changes involving three variables requires a simple strategy, but one exceedingly rich with implications and complications. An ordinary differential equation of the sort previously examined calculates change rates between two variables, one dependent and one independent. Other features in the equation are called constants and ever since Descartes's time have been

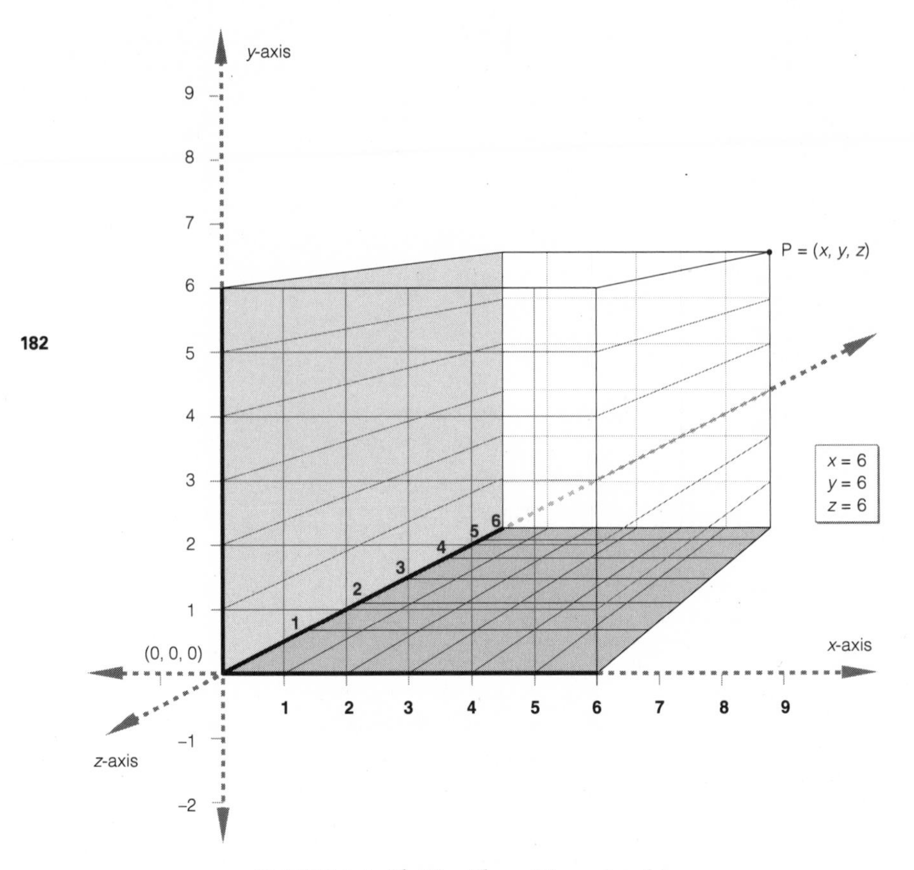

FIGURE 7.1. Plotting Three-Dimensional Space

designated by letters from the start of the alphabet. As fixed values, constants have no bearing on the changing relation between variables in the process of differentiation. This fact revealed to Lagrange and others a way to introduce numerous technical innovations in calculating.

The strategy itself builds on the idea of having symbols correspond to procedures. Where there are three variables in an equation (say, *x*, *y*, and *z*), simply consider one of them as a constant, then differentiate the other two, as we did with the tree case. This produces the "partial" derivative or differential equation. Next, repeat the procedure, each time treating a different variable as constant, until all the variables have been differentiated with respect to one another. If there are more than three variables, we can differentiate them two at a time, holding the remaining ones constant during the procedure. In a few cases the total differentiation of the original equation results from combining all the partial derivatives. Most often, however, the partial derivatives themselves suffice to calculate the change ratios sought.

Partial differentiation provides a simple calculating strategy, but one far-reaching in practice. It leads to an extensive, oftentimes taxing array of permutations, variations, applications, and consequences. To connect partial differential equations with physical laws generally involves second-order derivatives (sometimes rhetorically interpreted as "rates of change of rates of change of position"), among the most important of which for mathematical physics are the "Laplace equations" in two and three dimensions. We can sidestep further technical detail here and simply note that, much as with Lagrange's innovations, Laplace devised his generalized, partial differential equations from the techniques of manipulating ordinary derivatives containing constants. Then by letting the constants stand for other functions, Laplace created ways of combining various functions in order to map the specific motions of celestial bodies. His five volumes of celestial mechanics demonstrated these techniques (and his genius) for more than two thousand pages, required over a quarter century for completion, and became the most extensive exemplar of scientific determinism.

With such procedures not only could variables of multiple dimensions be calculated, but information could also be combined and manipulated with a hitherto undreamt flexibility. One brief example will illustrate. A vibrating string can be shown to satisfy a second-order partial differential equation: $\partial^2 y/\partial t^2 = c^2(\partial^2 y/\partial x^2)$, where t represents time, x and y are coordinates of a point on the string, and the symbol ∂ denotes the procedure of partial differentiation. The physical condition of the string is represented in the equation by a constant, the letter c. With a string of constant tension, c^2 = tension/density of mass.

But the condition of the string itself may change. Either it may be stretched tighter, altering the tension, or it may be composed of different materials, altering the density of its mass. Thus the information held fixed for the purposes of calculating a particular string's vibration at a given point in time may be changed in a subsequent calculation (say, as any piano tuner knows, on a less humid day). In this way information can be considered either as unchanging or as varying, according to circumstances. Herein lies the expansive scope of partial differential equations. They can take the laws of nature (the invariant portion of our information) and apply them to extremely variable, specific conditions; they are thus the engineer's instruments, par excellence.

Partial derivatives do this by setting (as they must) "initial" or "boundary" conditions for their calculations. These conditions comprise specific parameters of information—numerical or constant values of time, place, mass, materials, temperatures, pressures, and so on. In turn, constants are assigned the function of specifying the circumstances or constraints within which the calculating procedures of partial differentiation occur (solving equations for the variables). In the above example the string's physical composition (the constant c^2) serves as a boundary condition for calculating its vibrations. Additional boundary conditions might include specifying where the string "attaches" to the x-axis and how far away from the x-axis it "stretches" at t_0 (the starting point, with zero velocity). From such information and with partial derivatives, one can determine the location or velocity of any given point on the string.

The same holds true for many types of dynamical systems, those systems treating material bodies in motion under the influences of forces—such as a pendulum, a pump, a combustion engine, fluids and gases, a solar system, and so on. For a dynamical system evolving over time, one can begin with information about it at a specific moment, then map its evolution at successive instants. In turn, this mapping can often be further reduced and plotted, chartlike, on a coordinate grid of only two dimensions, then represented by equations of yet greater generality. Thus the flexibility of symbols, and in particular their capacity to correlate with calculating procedures and not just objects, can map useful information about our Heraclitean, ever-in-flux universe, whose "stuff" need not be assumed as fixed.

Sir William Rowan Hamilton devised an even more general formulation of the foregoing. Friend of Wordsworth and kindred romantic soul, Hamilton's imagination soared like a lark ascending into the ethereal abstractions of mathematical physics. His "General Method in Dynamics" essayed the pure mathematization of physical phenomena, of space, quantity, and matter. In fact, along with Lagrange and Laplace, he was d'Alembert's successor in this endeavor and can be said to have completed a century-long project of rationalizing Newtonian optics and mechanics by means of a rigorous mapping of nature's information.

Hamilton's achievement stands well outside our ability to comprehend it with image-laden rhetoric. It revolves around what has come to be called "phase space" (otherwise termed "state" or sometimes "configuration" space), an abstract, never-never land of unlimited dimensions (rather like Los Angeles). Phase space ranges far beyond the three-dimensionality (or even, with time, four-dimensionality) of our perceived experience. Nevertheless, by means of mathematical mapping or correspondence, it both permits storage of tremendous amounts of information about any dynamical system and generates powerful, predictive calculations about a dynamical system's future states.

In phase space complete knowledge about a dynamical system, or any particle in it, at any single instant, "collapses" into a point, as it were, and can be denoted by a single number. The point represents the mathematization of a physical particle. Introduced to dynamics by Newton, the term 'particle' in Hamilton's day bore its modern connotation as a hypothetical entity that possesses mass and observable position in space and time, but negligible size, having no physical extension. Matter is composed of such particles (twentieth-century atomic and subatomic particles are . . . another matter). It is rigid when the distance between them does not change with time; otherwise it is fluid, as with liquids and gases. When a particle is subjected to forces, the forces are considered to act at a single point, which in rigid bodies is the "center of mass," or "center of gravity."

Intuitively we can grasp how the location or position of any particle-point may correspond to the three axes of space and hence may be mapped onto a coordinate system as (x,y,z). Likewise we can imagine, at least somewhat, a three-coordinate system that maps the point's momentum—its impetus (technically, mass times velocity) as it proceeds in a particular direction. Yet Hamilton went much further and mapped these two systems into an unimaginable "space" of six dimensions in order to capture every aspect of a dynamical system. In his analysis, each particle has three momentum coordinates and three position coordinates, which are expressed in a single equation. Hence, for any number n of particles in the dynamical system, the space posited is 6 times n ($6n$) dimensions. This is "phase space."

Hamilton's highly reductive and abstract formulation enabled him to depict the total energy, kinetic and potential, of a dynamical system by means of a single mathematical equation, which he called a "characteristic function." The characteristic function (now written conventionally as H, a "Hamiltonian function") completely describes the dynamical system and expresses all the position and momentum variables of the particles within it. The partial differential equations that he derived from the characteristic function diagram in effect how the particles move through phase space in time; they thus provide a vector "picture" for the evolution of the system. This works for a rigid body and, more important, for a "vector field," in which the particles of a dynamical, fluid system operate. When a system changes, the phase points move to a new position; as it changes continuously, the points trace a trajectory.

The history of any dynamical system then can be read as a curve or orbit through phase space. Dense material this, whose access requires learning and thinking in a highly abstract language. Yet with it Hamilton completed the mathematical rationalization of Newtonian physics and even, we should add, helped pave the way for the revolution in quantum mechanics.

We have come about as far as we need or dare to in this brief sketch of the abstractions of mathematical physics. Elaborately constructed in the symbols of partial and ordinary differential equations and other functions of higher mathematics, these abstractions far surpassed the analytical vision of d'Alembert and the encyclopedists. With them scientists in the nineteenth century could capture vast galaxies of information about both the invariant and the fluctuating features of the natural (and, some still believed, social) universe and could generate precise and accurate predictions about its future states.

At the same time the scientists had also created their own attenuated and abstract universe populated with "entities" whose description defies sense, logically or metaphorically, in ordinary language. How could one speak with a straight face about a hypothetical "particle" of mass that had no extension? By definition 'extensionless mass' stands as a contradiction in terms (mass describes the gravitational tug of a material body, which extends and thus occupies space; a particle-point has no extension, thus occupies no space). The self-contradictory "particle" is simply postulated, as is the equally oxymoronic "phase space" in which it floats. They do not "exist" any more than does the grin of Lewis Carroll's Cheshire Cat.

We must take care, however, not to be too Hellenic here. True enough, nineteenth-century science became ever more replete with these rhetorical contradictions, the heirs to Berkeley's ghosts of departed quantities. But the information harbored and manipulated by these intricately constructed abstractions also brought countless operational successes in predicting phenomena—and has continued to do so. The inferences drawn from the successes were many and greatly buttressed the age's general confidence in scientific rationality. In the face of this confidence, the claim that the world cannot "be" or "exist" in the way described by the abstractions of mathematical physics demonstrated only the inadequacy of natural language in representing information. Symbols worked; words did not and could not—at least not in the same way and not to the same extent.

Nevertheless, the growing awareness of the inadequacy of words, combined with the relentless attenuations of mathematics and physics, led some mathematicians and logicians to concentrate on the logical principles underlying both language and analysis. For thinkers like De Morgan and Boole, among others, d'Alembert's naive faith in "leaping the gap" between phantoms and realities could no longer adequately assure the connection between the two. Both men believed that new techniques governing the construction and management of logical symbols were needed. Rigorously defined and properly understood, these techniques would tender a renewed and more viable foundation for the analytical vision, and a means of connecting it to our ordinary world of experience. As we shall see, the techniques were so good that, ironically, they uprooted the very vision that created them and then assumed an independent life in our own information age.

Language as Symbolic Analysis: De Morgan and Boole

With the hindsight of over three centuries' perspective on the emergence of symbolic mathematical thinking in the Western world, it seems odd to us at first glance that logic remained for so long thoroughly rooted in the classificatory tradition laid down by Aristotle. Why was this discipline of devising rigorous rules of inference and argument apparently so immune to the developments of symbolization in mathematics? Part of the reason lay with what scholars now term the "intuitive" nature of modern mathematics prior to the nineteenth century.

Until then observers and practitioners alike had generally followed the encyclopedists' account of mathematics as the "science of magnitude." Through the principles of correspondence and recurrence, magnitude embodied the intuitions of number and extension, of algebra and three-dimensional space, which were captured in abstract symbols and manipulated by the analytical tools of calculus. The intuitions of mathematics thus stood radically apart from the intuitions of terms, those "mental extracts" that made up the nouns and adjectives of natural language. Aristotle's logic had rested on terms, and his system of classification had arranged them. Both terms and the logical relations between them, he assumed, reflected things of the real world and the rational structures tying them together. On this assumption the study of logic stayed effectively outside mathematics for centuries, oblivious to Cartesian and Leibnizian sweet dreams of symbolic reason.

The situation changed dramatically during the nineteenth century, as two developments brought logic and mathematics together. First, the attenuation of the analytical vision led to the growing formalization of mathematics itself. Formalization entailed recasting many mathematical definitions and operations into a more rigorously defined, logical and symbolic form. Mathematics was becoming, as noted earlier, a postulational-deductive science. Second, the process of formalizing mathematics contributed to greater flexibility in the use of abstract symbols. This process freed symbols from the confines of "intuitive" mathematics and paved the way for their use in expressing natural language and the relations of logic. Together these two developments encouraged the search for the ultimate foundations of knowledge in the laws of symbolic reasoning, a quest that would eventually prove illusory.

Logic and mathematics began to merge through the growing formalization of mathematics in the 1820s and '30s, when British algebraists took an intense and renewed interest in the foundations of their discipline. Earlier attempts to devise and apply general theorems about algebraic equations had revealed difficulties surrounding negative and imaginary magnitudes. For example, to be general a theorem of algebra had to allow the substitution of any numerical magnitudes for the letters in an equation. This requirement led to the subtraction of a larger magnitude from a smaller one. But because magnitude was defined as always positive, a "negative magnitude" became another of those oxymoronic contradictions in terms. It made no more sense than had "instantaneous rate of change" a century before. (Equally nonsensical were the imaginary numbers, the square roots of negative numbers—absurd, because a negative number times a negative number always yields a positive number.) By this time, of course, mathematicians employed such terms with abandon, but the meaning of these "impossible" algebraic quantities remained unresolved.

One attack on the question was devised by the mathematician George Peacock (1791–1858), who argued that algebra simply manipulated abstract symbols according to prescribed rules, and that the symbols could be given any interpretation whatsoever as long as they were consistent with the rules. He thus etched a distinction between "arithmetical" and "symbolical algebra." With the former, algebraic symbols must stand for arithmetical quantities, whereas in symbolical algebra the symbols can have any meaning one might wish to assign them. Symbolical algebra, he added, does not even deal with numbers or magnitudes at all. True, its rules may be suggested by the rules of arithmetical algebra, but they do not depend logically upon them. Symbolical algebra comprises a separate science in which "impossible quantities" may be used freely, provided only they consistently follow the definitions and rules postulated at the outset.

The merging of mathematics and logic continued with Augustus De Morgan, whose *Formal Logic* (1849) challenged the traditional bases of Aristotelian logic. Logic concerns only names or words, De Morgan insisted, not ideas or forms or categories or any other mental extracts elicited from real objects, as commonly held. Moreover, words can be represented by symbols such as letters, as in 'All *S* is *P*', where *S* and *P* can stand for words corresponding to any objects or groups of objects. Symbols for words can be connected together, in turn, by other symbols that define the relations between them.

Here was the link to mathematical formalists, like Peacock, who believed that an algebraic law, such as the commutative principle '$a + b = b + a$', need not represent quantities or magnitudes. Any names or words or objects whatsoever could be substituted for *a* and *b*, provided only that they satisfied the law. De Morgan went further and claimed that the plus sign itself did not have to designate traditional addition. One could assign it a general meaning like "tied to" and then express many ordinary sentences completely with logical symbols. For example, the rhetorical statement 'if *a* is tied to *b*, then *b* is tied to *a*', becomes expressed symbolically as '$a + b = b + a$'. Moreover, by interpreting the plus symbol in this fashion it could stand for 'is', the critical copula of traditional logic. The plus sign or 'is' now could signify any type of connection between *S* and *P*, as long as the connection satisfied certain rules, including, for example: (1) the noncontradiction rule, where of two propositions, '*S* is *P*' and '*S* is not *P*', only one can be true; (2) the transitive rule, where if '*S* is *P*' and '*P* is *R*', then '*S* is *R*'.

De Morgan's approach established the general idea that different relations can be defined by arbitrarily assigning them symbols and by stipulating the symbols' functional properties. He later expanded his theory of relations and made the foregoing more extensive, rigorous, and precise. And even though he stayed generally within the Aristotelian framework, he rightly saw his work as a major departure from tradition. "For the first time in the history of knowledge," he pronounced, the logical "notion[s] of relation and *relation of relation* are symbolized."

De Morgan's new insights contributed directly to the turn from rhetorical to symbolic logic, a metamorphosis that parallels the sixteenth- and seventeenth-century shift in mathematics from rhetorical to symbolic algebra manifest in the achievements of Viète and Descartes. But if De Morgan was the Viète of symbolic logic, George Boole was its Descartes. In the opening pages of his seminal masterwork, *An Investigation of the Laws of Thought* (1854), Boole announced his capacious project, one surely of brazen, Cartesian dimensions. He would express the "fundamental laws" of mind in the "symbolical language of a Calculus," and he would thereby "establish the science of Logic and construct its method."

As the commonsense intuitions grounding both "language and number" (those "instrumental aids to the processes of reasoning," Boole called them) fell by the wayside, Boole became all the more convinced that his new logic would reunify knowledge through the laws of the symbol. Like Descartes, he too harbored a yearning for order. And also like Descartes, in his attempts to base this order upon "the foundation of a few simple axioms" of symbolic logic, he would unwittingly sever his technical innovations from his deepest wish. For now, we must keep in mind Boole's intention as we turn briefly to his logic. Just as we did with Descartes, we must resist the temptation to jump ahead into the implications of Boole's brilliant, novel techniques, lest we forget the conceit that midwifed them: his conviction that he had found the "fundamental laws of those operations of the mind by which reasoning is performed."

Boole's accomplishment could scarcely have been forecast from his upbringing. The son of an unsuccessful shopkeeper, he had battled poverty, social snobbery, and their attendant educational deprivations throughout his youth and early working life. But these same struggles had revealed an extraordinary drive and tenacity, which served him well throughout the life of his equally extraordinary intellect. An autodidact in the best, rarely applicable sense of the word, Boole was able to approach logic and mathematics with something of a fresh perspective. "This is the work of a real *thinker,*" Babbage wrote in his copy of Boole's first book, *The Mathematical Analysis of Logic.* Gradually he became known among a small circle of British intellectuals that included De Morgan, Peacock, and others doing innovative work in mathematical logic. Eventually too the Royal Society recognized his work, and with De Morgan's help, Boole obtained an academic post in Ireland. This freed him from chronic financial worries and allowed him to devote the rest of his life to teaching and writing.

In carrying out his project, Boole believed that he was uncovering "certain general principles founded in the very nature of language, by which the use of symbols, which are but the elements of scientific language, is determined." Thus would natural and symbolic languages be understood as thoroughly interlaced with one another and resting on the same postulates. The postulates or elements, he noted, were conventional and to that extent changeable and arbitrary. But once having "fixed" the meaning of the symbols used in our reasoning processes, "we never, in the same process of reasoning, depart." With these quaint expressions Boole declared his conviction that although symbols may be applied arbitrarily, the laws governing the relations between them are not. They are fixed. They are the "ultimate laws of thought" itself.

Boole's fresh perspective let him turn Aristotelian logic end for end. That logic had gravitated around words and things, proceeding from the "bottom up" first by formulating terms, then by eliciting the webs of relations between them. By contrast, Boolean logic began not with words and things, but with abstract symbols and the question of how they could be assigned to correspond with things. As was happening with mathematics during the nineteenth century, formal logic would become in Boole's hands a purely abstract algebra, a postulational-deductive science. We can see his project as emerging in two phases. First, Boole mapped words onto symbols and invented the rules for manipulating them; second, from these symbols and rules, he devised an operational, logical algebra of enormous breadth and flexibility.

Echoing others before him, including d'Alembert, Boole considered "abstraction" to be a "mental operation," a moving "process" rather than a fixed, conceptual object. But whereas d'Alembert believed the operation mapped a real universe, Boole took a different approach. For him, the "first step" of abstraction is the arbitrary "conception of the Universe itself," by which phrase he meant an infinite universe of conceivable objects, an infinite world of possibilities. (The sheer grandeur—and hubris—of our "imagination," which allows us to postulate arbitrarily such an open-ended universe, would eventually undercut Boole's conception of universal class.)

From the first step we take a second, an operation that requires "selection" (from the universe we have imagined) of some individuals "according to a prescribed principle or idea." Like the first, this process too can be totally arbitrary in that the principle for selection needs no particular grounding or foundation in preexisting objects, entities, or realities of any sort. The selection establishes a more restricted, "prescribed universe of discourse" and permits the fixing of our attention upon the member individuals within it. Together these two steps mark the "definite act of conception," the formation of abstract "collections" or "classes of objects."

Once having conceived abstract classes or collections of objects, we next assign symbols to represent them. A symbol or sign is just an "arbitrary mark," Boole noted, but one to which we give a "fixed interpretation." (He also used the term "elective symbols" for these marks.) This establishes a convention of usage. "Let us then agree" that a "single letter" can represent any "class" or "collection of individuals," which can correlate with some name, say "men." Symbols, we may agree further, can represent classes with only a single member, those with some members, those with no members ("nothing" or "no beings," the null class), and the universal class ("universe" or "all beings"). For Boole all natural language substantives (nouns and adjectives) can be represented as abstract classes, with each class being designated by a single, arbitrary letter.

Further, all verbs can be considered as the "substantive verb *is* or *are*" and represented by the equals sign of identity. (In this he diverged from De Morgan, who had considered the plus sign, among others, as capturing the copula 'is'.) Thus, 'the dogs are running' can be expressed symbolically as $x = r$, which states that the class of 'dog beings' (assigned the letter x) equals, or is the same as, the class of 'running beings' (call it r). Now, the equals sign between two class symbols means simply that the classes have the same members.

The next, critical twist lay with "forming the aggregate conception of a group of objects." The abstracting process yields arbitrary selections or groupings of objects from a universe of possibilities, to which we correlate names and assign letters. But to convert these abstract collections into effective instruments of thought, Boole saw that they must be manipulable, capable of being combined into new and different classes. Hence he constructed rules for aggregating classes in three basic ways, which would later be called logical addition (or union), logical multiplication (or intersection), and logical complementation (or negation). Additionally, the new, aggregated classes possess member subsets or subclasses, which could also be manipulated. With "algebraic" rules governing the operations of sets and subsets, Boole's logic would far supersede both the fixed classifications of Aristotle and the analytical reductionism conceived by the encyclopedists; it would provide unprecedented flexibility for manipulating abstract classes and the information they contained.

The rules themselves are straightforward and uncomplicated. In logical addition, the sum or union of any two collections, x and y, is simply the newly formed class or set containing all the members that belong to either x or y, or to both x and y. Where the collections overlap, their members are counted only once, so as to avoid redundancy. Letting x stand for the class of dogs and y for that of cats, $x \cup y$ represents the larger class that includes them both, noted by the letter W, so that $x \cup y = W$. (Today the symbols $\oplus$ or $\cup$ generally connote logical union.)

The terms of logical addition obey many of the same rules as the numbers of arithmetic, such as the commutability principle. Logically, $x \cup y = y \cup x$ because 'dogs and cats' says the same thing as 'cats and dogs'. Whereas the plus ($\oplus$ or $\cup$) designates the process of logical addition, the minus ($-$) for Boole refers to the exception or partition of classes. For example, if W means the universal class of all dogs and cats, and zx means short-haired dogs, then '$W - zx$' means 'the universal class of all dogs and cats, excluding short-haired dogs'.

Classes can also be combined through the operations of logical multiplication or intersection, as we just did by creating the class of short-haired dogs (zx). The intersection of two classes entails aggregating into a new class only those members common to both of them. With x designating dogs, we let z stand for the class of short-haired beings. Then zx (or '$z \cap x$') stands for short-haired dogs. (The symbols ⊗ or ∩ are used for logical multiplication, but as with algebraic multiplying they are frequently omitted.) The intersection of dogs and short-haired beings represents a subset of two collections, dogs and short-haired beings, a subset whose members have something in common, namely that they are *both* dogs and short-haired.

As with logical addition, many manipulations of logical multiplication resemble those of numerical algebra. If zx = short-haired dogs and zy = short-haired cats, as a case in point, our saying 'short-haired dogs and cats', $z(x \cup y)$, clearly means the same as saying 'short-haired dogs and short-haired cats', or $zx \cup zy$. Therefore, $z(x \cup y) = zx \cup zy$. This "principle of distribution" holds true also for numerical algebra: $z(x + y) = zx + zy$.

Logical complementation, or negation, was the third of Boole's basic operations for manipulating abstract classes. To understand it requires introducing one of his most fertile insights, namely, that the number 1 can represent the entire, imagined universe or, he later amended, a prescribed universe of discourse, while the symbol 0 can represent nothing—the empty or null set. We let x represent any class of objects, such as all dogs. From our conception of the universe, now denoted by the number 1, we then remove x, so that there remain all the other members of the universe, excluding dogs. This remainder is the class of all nondogs, the complement of the dog class, and is symbolically expressed as $1 - x$, or the abbreviated symbol x'. It follows that if x itself were the entire, prescribed universe of discourse, 1, its complement (x') would be 0. All or nothing, therefore, provide the fundamental, logical complements from which we select other member classes and their complements in the course of reasoning.

The logical operations of addition, multiplication, and complementation lay at the heart of Boole's laws of the symbol and served as the starting point for deriving many other, far more complicated formulas to be used in manipulating abstract classes. In the next chapter we shall see how these operations would be utilized to great effect in the techniques of computer hardware and software design. Yet for Boole, we must remember, the operations signified much more than the techniques of information management. Anchored in classes and in the simple rules governing their combination, logical operations revealed the foundations of all thought.

Boole expanded his operations of symbols into a full-fledged logical algebra, which would eventually complete the assimilation of natural language into a purely symbolic universe. This expansion led to what is now generally called the propositional or sentential calculus of logic, effectively begun by Boole but completed by subsequent generations of logicians. (He termed his own work "a Calculus," whose grasp required "acquaintance with the principles of Algebra.") In the calculus of propositions or sentences, symbols designate whole statements and the logical operations between them, in addition to standing for classes or collections of objects. And just as the symbols of mathematical calculus can correspond to calculating procedures (and not merely to collections of units and the fixed relations between them), so too can logical calculus be made to correlate with linguistic operations. Boole's logical algebra thus launched the transformation of natural language syntax into a flexible, rule-governed instrument for describing the world with precision.

Despite their obvious kinship, Boole distinguished his logical algebra from traditional, numerical algebra with great care, for not all the arithmetic functions of the latter carried over directly. With minor exceptions, for example, he ruled out division. More important, consider what happens when classes are added to or multiplied by themselves. Unlike numerical operations, in logic when a class combines with itself, it produces only itself: $x \cup x = x$, not $2x$; $xx = x$, not x^2. These operations say simply that if we aggregate all members of the dog class with all members of the dog class, we shall still have only all the dogs, not twice the number of dogs or the number squared. This feature of logical combinations, which Boole termed the "index law," now goes by the name of the "idempotent" rule. It states that all the powers of a set are the same, that $x \cup x = x$, $x \cap x = x^2 = x$, $xxx = (x^2) \cap (x) = xx = x$, and so on.

With the idea of idempotency Boole broached his most intriguing and far-reaching innovation, the demonstration that logical algebra can hold for two numbers (1 and 0), which correspond respectively to the universal and null sets, as noted above. He first showed how this works by means of simple substitutions. Given the relation $x^2 = x$, he observed, two numerical possibilities alone can satisfy the equation. These are 1 and 0, for $1^2 = 1$ and $0^2 = 0$. Other logical operations with 1 and 0 also match those of numerical algebra ($1 \cup 0 = 1$; $0 \cup 0 = 0$; $1 \cap 0 = 0$), except for idempotent addition, where $1 \cup 1 = 1$. By themselves these exemplars of logical operations do not appear startling. But then Boole applied the foregoing to the truth and falsity of propositions. We can use x to denote "the proposition X is true," he wrote, and we can let 1 refer to the time frame during which x applies, so that $x = 1$. By these means we capture symbolically the truth claim of a proposition. In a similar fashion, we can employ $x = 0$ to represent the statement "the proposition X is false." In making these assignations, he implicitly added a new, exclusionary principle to his analysis, namely that either $x = 1$ or $x = 0$, *but not both.* This principle transformed his rules of aggregation and complementation into a bona fide "two-valued algebra" of classes. It metamorphosed 1 and 0 into logical operators.

TABLE 7.1
Boolean Two-Valued Algebra

Addition (or union)			Multiplication (or intersection)			Complementation (or negation)	
∪	0	1	∩	0	1	W	W'
0	0	1	0	0	0	0	1
1	1	1	1	0	1	1	0

The two-valued algebra opened up boundless technical possibilities in subsequent applications by establishing the matrices of logical relations between 1 and 0. We can demonstrate these matrices with a simple, two-place universe, consisting of a set with a single element and its complement. To do so, we let $W = 1 = \{x\}$ and its complement $W' = 0 = \{\ \}$. This produces a binary world in which the logical operations in table 7.1 hold. Reading across by row and down by column, these tables present all the possibilities of the Boolean operations of logical addition, multiplication, and complementation.

When imposed on propositions, the above tables become logical operators, transforming the symbols of Boolean algebra into the propositional calculus. A proposition involves more than an abstract class with members. As a declarative statement it expresses an idea about those members, an idea that can be considered either true or false. Otherwise stated, propositions possess "truth-value," a term later coined by logician Gottlob Frege (1848–1925). Thus, 'some cats are short-haired' and 'all small dogs are high-strung' are propositions whose truth or falsity can be assigned or established. We do so in these instances empirically, by seeing some short-haired cats, or by never finding a small, mellow dog.

The rules governing the relations between propositions and their accompanying truth-values comprise the logical connectives of 'and', 'or', and 'not'. They derive directly from Boole's two-valued algebra and are expressed by the tables. To see how, let us begin with 'not', or negation. We let p stand for any proposition, such as 'some dogs are short-haired' and $\sim p$ for its negation, 'no dogs are short-haired'. (Logicians commonly express propositions by lowercase letters, such as p, q, and r, which are variables that can stand for any statements whatsoever.) The following table expresses the possibilities in logical negation:

p	$\sim p$
F(0)	T(1)
T(1)	F(0)

To interpret the table, we let the letters T and F designate the truth or falsity we may assign to any proposition, while 0 and 1 indicate the correlation with the complementation table in the two-valued universe of classes. If p is false (0), then its negative $\sim p$ is true (1), and vice versa. In fact, for logicians the truth table itself has become the means of defining negation.

In the same manner other Boolean tables can be devised to express the truth values accompanying the logical relations of conjunction ('and') and disjunction ('or') between propositions. Note that conjunction is not the same as logical addition or union (as our linguistic intuition would have us believe); rather it corresponds to multiplication, or intersection. Likewise disjunction is not multiplication or intersection, but rather correlates with addition, or union (see table 7.2). Here, the truth table for conjunction, whose logical symbol is $\wedge$, matches that of Boolean multiplication or intersection, represented by the 1s and 0s in parentheses. A conjunction is a compound statement linking two propositions; as the table shows, if both components (or conjuncts) are true, then the whole conjunction will be true. If p is true and q is true, we read, $p \wedge q$ is also true. Otherwise the conjunction will be false.

TABLE 7.2
Boolean Rules as Logical Operators

	Conjunction (AND)			Disjunction (OR)		
		(Truth value of q)			(Truth value of q)	
	$\wedge$	$F(0)$	$T(1)$	$\vee$	$F(0)$	$T(1)$
(Truth value of p)	$F(0)$	$F(0)$	$F(0)$	$F(0)$	$F(0)$	$T(1)$
	$T(1)$	$F(0)$	$T(1)$	$T(1)$	$T(1)$	$T(1)$

Likewise with disjunction, which generally carries the logical symbol $\vee$. From the table we see that if a proposition p is false and another proposition q is false, then their disjunction $p \vee q$ will be false. In all other cases, the entire disjunctive proposition $p \vee q$ remains logically true because at least one of the disjuncts is true.

In the foregoing tables, the 1s and 0s reveal the mapping of conjunction, disjunction, and negation with the earlier lattices of multiplication, addition, and complementation. In all three cases, rather than having the symbols 1 and 0 correlate one-to-one with fixed, abstract classes of things, they now serve as functional, logical operators (whence the contemporary "Boolean operator"). They "operate" in the propositional calculus by assigning truth-values to propositions in accordance with the schema of logical rules Boole had devised. Their role here suggests the most profound implication of Boole's two-valued algebra: the unlimited algorithmic possibilities of symbolic logic for framing (in the earlier-cited words of Norbert Wiener) the "content" of our exchanges with the outer world, for framing information.

Boole's innovations, like those of Descartes, had enormous consequences for his successors, far beyond his own imaginings. Looking ahead, one practical implication lay with the possibility of integrating his two-place algebra of classes with a binary number system capable of performing any of the algorithmic calculations of higher mathematics. From Boolean beginnings, "logical algebra" and "numerical algebra" would eventually merge in the hardwares and softwares of the successors to Babbage's analytical engine and become the driving conceptual force behind the modern computer. Similarly, in the hands of later logicians, the propositional calculus would quickly supersede its Boolean origins, developing the far more sophisticated techniques of quantification and argumentation that have come to typify contemporary symbolic logic.

Gone from this new logic were the traditional assumptions about words and things that had underlain the developments in alphabetic literacy and that had culminated in the Aristotelian, classificatory way of informing the world. No substances (metaphysical or otherwise) lurked in Boole's abstract classes. Rather, they served as empty containers for any information whatsoever. This was a whole new conception of how words and things relate through symbols. We still retain the general idea of classifying things, but now it enters from afar, through the analytical door so to speak, and stands enmeshed in the pure techniques of symbol formation and manipulation.

Henceforth, information can be organized in classes, but the classes are subject to liberations and constraints previously unknown and scarcely imagined. Take liberations first. Without the presumption of a substance or reality in nature that imposes itself onto one's categories, the grouping of sets, subsets, and members becomes purely arbitrary. Classifying depends entirely on conventions agreed upon, or desired, by the creators of classes. By implication, new conventions of classifying can be formulated and agreed upon as well. The arrival of the computer will enable its users to devise categories as haphazardly or restrictedly as they wish. Classifying will become play.

Second, the constraints. Although conventional, the basic techniques for creating, connecting, and manipulating abstract classes are at bottom like the postulated rules that serve to underlie the operations and procedures of abstract mathematics. We need to keep in mind that symbolic Boolean logic is an algebra, even though its rules differ somewhat from other numerical or abstract algebras. The constraints will therefore depend on the specific rules assigned for manipulating informational symbols. Some rules work better than others, so well, in fact, that they apply to a broad range of phenomena and tasks. In these cases the constraints of rule-bound systems actually open up new vistas of information processing and transform the computer into an instrument of knowledge discovery, as well as one of knowledge organization.

The Erosion of Analytical Foundations

The nineteenth century displayed unbounded confidence in the analytical vision. Yet in the early decades of the twentieth century, the abstractions and technical accomplishments of both mathematical physics and symbolic logic combined to undermine even the most assured foundations on which these disciplines rested. This erosion and eventual uprooting of analysis exposed the irony of the nineteenth century's conceit. Here, we can only allude to these developments with passing mention of two profound challenges to analysis.

First, in the early decades of the twentieth century the discovery of quantum physics revealed empirical limits to the determinism of scientific abstractions. In particular, the so-called "quantum weirdness" that characterizes the behavior of subatomic particles lies outside the linear and deterministic predictions of classical physics, such as those devised by Hamilton. This behavior has been confirmed in experiments, which have shown that one cannot map simultaneously both the momentum and position of any given subatomic particle, an electron for example. As a consequence, summarizes physicist Heinz Pagels, "the smallest wheels of the great clockwork, the atoms, do not obey deterministic laws." As a further consequence, the reductionism of twentieth-century physical science reveals its own limits.

The second challenge to the analytical vision came from the inside, so to speak. Two discoveries in particular illustrate the conundrum of analytical reason turning on itself. The first occurred in the form of an epistolary bomb, written in 1901 by Bertrand Russell to fellow logician Gottlob Frege. In the letter Russell announced discovery of the famous "class paradox" that now bears his name. He demonstrated that the abstract idea of universal class was self-contradictory when pushed to its logical extremes. The result is like a variation of the liar's paradox, which arises from saying 'all general statements are false'. This statement is general and therefore false; or, if it is true it must be false; therefore it is false. To assert the statement is simultaneously to deny it; hence the paradox. Discovery of Russell's paradox struck a damaging blow to the concept of universal class (the "class in which are found *all* the individuals that exist in *any* class"), on which Boole had based his entire *Laws of Thought*.

Then in 1931 the brilliant, mathematical logician Kurt Gödel delivered a second, even greater blow to the foundations of analysis. He proved that no system of formal mathematics or logic founded on axioms could be at once complete, consistent, and self-contained. From Gödel's analysis it followed that any formal system of symbolic reasoning required postulates or propositions for which the system itself did not have proof. Complementing the empirical limits to reductionism discovered by physicists, the arguments of Russell, Gödel, and a host of others showed internal limits to the reductionism of logical analysis by undermining the universality of its conceptual bases.

These internal challenges to analysis were set up by the work of Boole himself and may be seen as the ironical consequence of his own hubris, grounded in genius. In the name of shoring up the "laws of thought," Boole had injected arbitrariness into the heart of the analytical vision—in imagining the universe, in conceiving general and specific classes, and in assigning symbols to classes. In effect, he believed that we could begin our thinking, arbitrarily, with whatever worlds we could imagine. The laws of symbolic thought would connect us to these worlds, for which we needed only the axiomatic concept of universal class and the formal rules combining abstract classes. Russell's and Gödel's discoveries laid bare the arbitrariness of Boole's foundationalist desires, and in so doing they revealed the irony hidden in his technical accomplishments. Intended to establish a new foundation for the analytical vision, Boole's techniques (and their successors) created instead a world all their own, one of symbol management severed from any necessary ties to the universe outside.

The uprooting of analysis brought the modern information age to its end. Flakey implications are often drawn from these developments, particularly when such complex arguments are caricatured, as the limits of space have imposed on us. So we need to make clear just what this uprooting signified. It did not, we stress, and does not mean an end to the accumulation and ever-growing sophistication of modern scientific learning. Nor does it mean that "anything goes" in the analytical temper of today's sciences, logics, or mathematics. Still less does it mean jettisoning the technical accomplishments of modernity into a cauldron of postmodern perspectives and enthusiasms. The information idiom of analysis continues to provide a useful, practical structure for molding much of our experience.

The consequences of "analysis uprooted" lie elsewhere. With the erosion of global, foundational claims—the cornerstone of the eighteenth century's analytical vision—the information promulgated by the technology of numeracy and the analytical idiom of knowledge could no longer provide an all-inclusive abstract model of the universe. This fact alone suggests profound limits on the means of closure to humankind's collective compulsion for gathering and arranging information.

In antiquity classification had provided closure by eliciting a hierarchical arrangement of categories from the nouns and adjectives of natural language and by correlating these words with collections of real objects. As classificatory closure broke down in the sixteenth century, another language, the abstract language of symbolic analysis, steadily inserted its own version of a closing hierarchy into the exhaustive process of information collection. This was the reductionism of operational symbols, manifest in relational mathematics, in the organized knowledge of the *Encyclopedia,* and later on in symbolic logic.

The uprooting of analysis in the twentieth century has undercut the means of closure found in both classification and analysis. A global uncertainty has replaced global foundations. But this loss of closure has not produced the same sort of intellectual crisis experienced by Montaigne and his contemporaries. Our age does not share John Donne's bewilderment and loss of orientation, even though we too are inundated with information. The reason for this lies with the unprecedented growth and expansion of our technical means of information mastery, the explosion of information technology that we call the computer revolution.

CHAPTER EIGHT

The Realm of Pure Technique

> Information is just signs and numbers, while knowledge has semantic value.
> —HEINZ PAGELS, *The Dreams of Reason*

Information in the Contemporary Age

No longer a universal, abstract model of the world, either classifying or analytical, information today has become a world unto itself. It is a world whose abstract symbols can be assigned arbitrarily to any objects and procedures whatsoever. Although symbols do not mirror extra-symbolic reality in any profound, philosophical manner, the technical operations manipulating them remain governed by the rules of a tightly knit logic. Symbols stand in bizarre conjunction—at once determinate and utterly capricious. Long ago, Boole himself had proclaimed this critical idea. Symbols functioned according to "laws" that were not arbitrary, he observed, whereas their interpretation was quite another matter—not law-governed, but arbitrarily dependent upon their user.

In the twentieth century, analytical symbols have found a new home in the electronic computer. Together, symbol and computer have joined in symbiotic union to create the information technology and idiom of our contemporary age. This epoch-defining innovation takes its place alongside those of alphabetic literacy and modern numeracy. Culminating with the alphabet, early writing had created an information technology of unprecedented scope, one that educed the classifying potential of natural language. The rise of relational mathematics in the modern age, likewise, had realized the information potential of number and organized it in a broad-reaching, reductionist hierarchy.

Today in turn, the computer has elicited the information potential of purely abstract symbols, fabricating a realm of pure technique apart from any foundation in knowledge.

This marriage of symbol and computer has given birth to the identifying hallmarks of our age: power and play. Central to the intellectual and cultural "ether of our times," as one observer labels today's information, these hallmarks issue directly from the determinacy and caprice that characterize abstract symbols. Power and play mark contemporary information not simply as a new idiom for conveying bygone wisdom and knowledge. Even less are they merely the adventitious by-products or personal possessions of information mongers and managers. Rather, they have come to constitute and define anew the portion of our experience that we call information. Strange as it may seem, they comprise the actual content or "stuff" of our exchanges with the world.

The power and play of information stem jointly from the logic that binds symbols and, as well, from the speed at which symbols are processed. Admitting no escape from its chains of reason, logic ineluctably creates information's power in the determinism of computer algorithms, the precise steps of information management. This determinism comprises power because each step in a computer program determines, and thus controls, each successor step. Applied to untold billions of data bits stored in electronic memory, logical procedures enable their user to command, control, and process entire universes of information at unimaginable, though not incalculable, speeds.

Information's power is bound intrinsically to play in two ways. First, the interpretation and application of logical symbols depend arbitrarily on the user, as Boole noted; no fixed, universal model governs the encoding of our experience into symbols. We create and process digital information in a manner quintessentially playful and capricious. Second, by allowing the iterations of the computer to run their logical course, we give new and often surprising, unpredictable shapes to information. The power of reiterating computer algorithms rapidly, and virtually endlessly, has created play in the modern age's structure of knowledge, loosening the grip of the reductive sciences and allowing new forms of knowing to emerge.

Viewed historically the rise of the new information idiom marks a stunning break with the past. Information in the classical age was thoroughly wedded to an idea of substance, as the nouns and adjectives of natural language had their semantic references in the world of things. In the modern age, d'Alembert's phantoms had reduced information substances to their ghostly apparitions in formulas, which connoted units of force, distance, mass, and so on, and which could then "leap the gap" to map the world outside. Even though language—

with its conceptual conundrums such as "instantaneous rate of change" or "phase space"—could not keep pace with the abstractions of analysis, the age itself eventually came to "make sense" as a world map governed by formulaic laws.

Now, in contrast to the previous ages, substance has vanished entirely from information. Our information technology stakes out a realm in which meaning or content—what earlier ages had abstracted from experience, shaped or formed as information, and understood as contained in memory—is replaced by logical rules. We have drawn the information idiom so far away from the immediacy of experience that no content whatsoever is retained in its digital symbols. The rules simply bracket, or delineate, the algorithmic procedures by means of which a piece of experience can be encoded and thus informed. Paraphrasing Montaigne from an earlier time, there is no bean in the digital cake. The technical processing of empty symbols, an activity governed entirely by power and play, has replaced all the beans, all vestiges of substance.

Nothing is fixed in today's information. The continuous flux of symbols collapses our usual distinction between information processing and its products. They are one and the same, mutually embodying power and play. Although not drawing this implication out of his own work, Boole had nonetheless captured its central insight. From his day forward, logic—that is, formal, symbolic logic—has been understood as the mutual entailment of procedure and relation. Logical relations (or rules, such as those displayed in Boole's tables) actually define and guide the procedures one follows in moving symbols about. Conversely, logical procedures themselves constitute the actual exercise of the relations or rules governing the movement of symbols. Writ large and embedded in the electronic computer, this mutual entailment produces the constant motion of symbols in accordance with logical rules. What we call a piece of fixed information represents only an arbitrary point at which we have halted the movement of information processing.

At its core, then, today's information strikes us as weird and wrenching. It rubs against the grain of common sense. Like much in our world, when we analyze information, break it down into its tiniest, most abstract (how else might we say it?) bits, it turns out to be empty, a void, form without content. Nothing "real" corresponds to a digital 1 or 0 (except perhaps, critically, ON or OFF). Yet, bound together in logical strings and operations these microcosmic bits begin to acquire patterned shape or form. As they do, information—understood as the encoded content or "stuff" of our exchanges with the world—emerges from rules and from the movement of symbols in accordance with them. It flows from the algorithmic manipulations of empty information symbols, from processing. Thus from digital bits does information beget itself. Its macrocosmic, final product at the level of what we see is but the result of positional changes of symbols. In linguistic terms, syntax, not semantics, governs the realm of pure technique.

With their mathematical and technical brilliance, two men in particular, Alan Turing (1912–54) and John von Neumann (1903–57), stand out as the principal architects of the new information age. Information's power, they both stressed, derives from yoking logical operations to electronic circuitry. Logic tenders the rules for moving symbols about in a "completely determined" manner (thus Turing), while circuitry corresponding to these rules actually manipulates symbolized information by directing electronic impulses through a "flow," or series of "control sequence" and/or "branching points" (thus von Neumann). Both men conceived the computer as electronically processing digitized symbols according to exact and undeviating steps.

Turing and von Neumann were equally concerned with the second hallmark of our information age, play. In fact, they were both fascinated with games in their personal as well as their professional lives. Turing, for instance, devised one of the earliest versions of computer chess, while the mathematical field of game theory owes its origins largely to the poker-playing von Neumann. And both toyed with the computer as a plaything in its own right. Their interest reveals deeper, more culturally rich levels of play, which connect the determinism of the computer's symbols and operations to the sensory world of experience.

Heralded by the accomplishments of Turing and von Neumann, power and play provide us historical purchase on contemporary information as pure technique. They stand to our computer-driven information technology as wisdom stood to literacy, and as knowledge to numeracy: the cultural and intellectual distillate of an information age. They permeate the micro-informational, digital universe within the computer, the 1s and 0s that comprise information at its most basic level. As well, they comprise the sensory-laden, macro-informational world, that portion of our experience captured in digital 1s and 0s. Even more to the point, through the magic of encoding, power and play frame the interchange between these two dimensions of information. They define our information age.

We shall begin exploring the realm of pure technique by concentrating in this chapter on the power and determinism of micro-information, the muscular progeny of symbol and computer. To do so, we must first look briefly at the stored-program, electronic computer itself in order to see how its material construction has combined with logic to produce today's information technology. For this purpose we need only minimal grasp of the machine's physical materials. Of more significance will be the machine's overall logic and design as devised by Turing and von Neumann. They conceived and defined the microcosmic universe of information's power. Understanding this universe will enable us to reconnect with the now digitized, macro-informational world of sensory experience as we return to information's play in the next chapter.

Electronic Information: The Computer

The creation and evolution of the computer have already filtered into the myth and lore of this heroic, larger-than-life technology. Its inventors scroll by like a pantheon of technical and scientific deities: Howard Aiken, John Atanasoff, Charles Babbage, Vannevar Bush, J. Presper Eckert Jr., Herman Hollerith, John Mauchly, Claude Shannon, George Stibitz, Alan Turing, John von Neumann, Norbert Wiener, among many others. In one way or another all these men contributed mightily to a remarkably simple idea. For in its essence, a computer is nothing more than some mechanical or electric or electronic (soon to be biological?) means of transferring information.

Before they were performed by inanimate machines, humans carried out information transfers, we can even say mechanically. Humans process information, move it around, make it proceed from one place to another, often according to instructional rules and maps provided. In their Greek origins, the terms for 'machine' and 'mechanism' (*mēchanē, mēchos*) referred to "means" or "expedients" and thus to the property of "having power" or "being able," which applied to humans and other animate beings, as well as inanimate objects. (Our word 'may' also derives from *mēchos* and maintains the sense of being able, or permitting.)

In the modern period, mechanization of thought long preceded the construction of "thinking" machines. Until quite recently the term 'computer' designated a person who computes, and in this sense Al-Khwarizmi's real achievement lay in discovering that one could devise rules for manipulating pieces of information in exactly parallel and equivalent ways, in an equation. When we compute or count we add information unit by unit (or, using only 1s and 0s, binary digit by binary digit, bit by bit). Conceptually speaking, humankind has taken a relatively small step from processing information with algorithmic mechanisms to the creation of information machines. Cybernetics inventor and early computer whiz Norbert Wiener (whom we met earlier) even noted once that because a computer is only a system of information transfers, the body of the machine could actually be constructed with any materials whatsoever. (One of his MIT colleagues built a simple working computer out of stones and toilet paper rolls to prove the point; California philosopher John Searle later suggested using empty beer cans.)

From the earliest simple calculators to contemporary computers of mammoth processing capabilities, two fundamental means of transferring and moving pieces of information around systematically have framed the development of all computational devices. These means define analogue and digital computers.

As their name suggests, analogue machines depend upon analogies in order to perform various computations. These analogies correlate some physical system with numbers and mathematical functions, thereby allowing their user to represent and calculate numbers mechanically or electronically. Another name for analogue devices is "continuous computers" (or sometimes "measurement computers"), which refers to the representation of numbers as continuous physical magnitudes, such as the lengths of rods, intensities of voltage, rotations of axes, wheels, gears, cams, and so on in various interconnected ways. Just as numbers can stand for physical quantities, like a stick's length, so too can physical quantities represent numbers. The slide rule is one familiar, analogue computer. It uses two sticks, marked off in graduated lengths, which correspond to the logarithms of natural numbers. Multiplication involves simple addition of the logarithms, achieved by sliding the sticks, matching the desired numbers, then reading the results at the marker. Similarly with other operations.

The mapping of continuous phenomena with analogue computers calls to mind our earlier discussion of the calculus and its affiliation with the ordinal concept of number. A differential equation designates calculating procedures whose logical end is rhetorical absurdity (the "instantaneous rate of change"), but whose practical end yields a simulation of change rates between dependent and independent variables, a simulation that can correlate to various phenomena. Its premise is the principle of recurrence. Having performed the procedure for a given numerical value, one can repeat the procedure again and again, as often as desired, substituting in each successive iteration a new number, $n + 1$, for the previous number, n. Whatever materials it uses, analogue devices embody this principle as they map continuous functions onto analogous mechanical, hydraulic, pneumatic, optical, electric, or electronic counterparts.

Within the realm of analogue computers, a division between specific and general machines sorts out those whose functions are confined to particular calculating purposes, such as the familiar automobile speedometer and electric wattage counter, from those whose computing functions admit a broader range of applications. Among the latter are general computers that can solve complex differential equations; the differential analyzers developed by MIT Professor Vannevar Bush in the 1920s and '30s were the first such machines. Still, however broad and powerful their applications, analogue devices perform almost exclusively as calculators, and their uses are restricted to doing with far greater facility what mathematicians could in theory do with pencil stub and paper. They remain halfway steps to the all-purpose digital machines that signify the breakthrough into realm of pure technique.

With the advent of electronic, digital computers we cross the threshold into the contemporary information age. The term 'digit' meant and still means finger (or toe). It also refers to the numerals used in counting, no doubt derived from the age-old practice of counting on one's fingers. If we think of counting on fingers discretely, one at a time, we have pictured a key feature of digital computers. In contrast to the continuity of the analogue devices and to their affiliation with the ordinal conception of number, digital computers are directly associated with the cardinal concept of number and are constructed around the one-to-one correspondences of discrete, individual units. A finger/unit stands for an object or thing or procedure, a symbol for a finger/unit, an electrical circuit for a symbol, and so on. Before the invention of computers, digital counting was generally performed in common, decimal numbers. (Digital thus does not mean binary, which refers to a base-2 number system, about which more below.) An abacus is a simple digital computer generally using a base-10 or base-5 number system.

When considered solely as a calculating device, digital machines can carry several distinct advantages over analogue computers. In the first instance, it is impossible to construct an analogue machine sufficiently general to calculate any mathematical function whatsoever. By contrast, digital machines can be completely universal, mainly because they perform only the four basic arithmetic operations. They add, subtract, multiply, and divide. More basic still, as d'Alembert well understood, products and quotients are built up of sums and differences. And because subtraction can be performed through a method known as "adding complements," one can actually construct a full-fledged digital calculator with only an "adder." Blaise Pascal had incorporated the complement principle in the first adding machine, which he invented long ago in the 1640s (see plate 14). To see how it functions, imagine a Boolean set with 100 elements, the whole numbers 1 through 100. The complement of, say, 36 is 64 (100 – 36). A subtraction such as 53 – 36 = 17 can be achieved by adding to the minuend, 53, the subtrahend's complement, which is 64, and deleting the hundreds digit in the difference: 53 + 64 = 017 = 17. Because the set cannot exceed 100, the hundreds digit, 1, returns to 0 with a carry, a technique utilized in "ring counters" like the automobile odometer. By implication, all mathematical computations, however complex, can be constructed digitally and piecemeal from the single operation of addition.

Moreover—a second advantage—digital computers can achieve greater accuracy and precision, as compared with their analogue counterparts. All analogue machines harbor a certain amount of vagueness, known technically as "noise," which describes the disturbing influences of the machine's physical materials on its calculations. The greater the precision in machinery or materials, of course, the less the noise. Yet, the "noise level" remains relative in analogue computers. Thus the accuracy of slide rules increases only marginally with the use of larger sticks. A simple, foot-long slide rule yields a precision of about 1 part in 300 (or 0.33 percent). To achieve a precision of 1 part in 1,000 (0.1 percent) requires sticks over thirty feet long and extremely elaborate construction. Even with greater precision, the graduated calibrations inevitably require some guesswork in computations, whereas in this sense digital machines are virtually "noiseless."

To picture this even more graphically observe the difference between a digital watch and an ordinary mechanical watch, which is in effect an analogue measuring device. The hands of the analogue watch hover over the dial marks and often appear to waver slightly in front or behind, depending on the angle of our vision in reading them and on the material quality of the watch. Extremely accurate timing suffers accordingly. The readout of a digital watch, however, is precise and unambiguous, with no mechanical sliding or shading from one number to the next. The same point holds for other digital devices.

A third advantage of digital computers resides with their facility for storing partial solutions and other results generated during the course of long calculations. Unlike analogue instruments, digital machines possess the capability of reserving discrete bits of information in available memory. These can be called upon and fed back into equations and computations as needed. Overall, then, measured by the single criterion of calculating—moving information according to mathematical algorithms and formulas—the digital computer offers distinct advantages over analogue machines: generality, accuracy, memory. Babbage had grasped this critical insight with his analytical engine, the first attempt at a general-purpose digital computer, occurring in an era when analogue devices were fast becoming the rage.

Despite their many advantages in principle, digital computers did not supplant analogue devices for nearly a hundred years after Babbage's invention, largely because they were so much slower. True, they are in principle far simpler, basing all calculations on the four functions of elementary arithmetic. Yet the very simplicity of digital machines comes at great cost, namely, having to go through manifold more steps to complete complicated calculations. Multiplying large numbers, for example, requires many discrete operations of adding and carrying, whereas with an analogue computer like the slide rule one can accomplish the same objective in far fewer steps. (For this reason, also, analogue machines continue to be used in many applications, such as guidance systems requiring automatic feedback and immediate, "real-time" adjustments.) Only with the arrival of electronic means of computation did the possibility of performing rapidly such a massive volume of calculations make digital computing more efficient and desirable.

In digital devices, therefore, the speed of calculations becomes critical. And with the technological developments of the past half century this speed has become mind numbing. Consider the following. A person multiplying a ten-digit number without any mechanical assistance normally needs about five minutes to complete the task. In the 1940s, adept human computers, aided by mechanical desk calculators, could multiply ten-digit numbers in about ten to fifteen seconds. One of the earliest digital computers, the "Harvard-IBM" machine designed by Howard Aiken and completed in 1943, required only three seconds for the same multiplication. Named officially the Automatic Sequence Controlled Calculator (ASCC), Aiken's machine used electromechanical relays and at the time produced extremely fast and reliable calculations. Yet it plodded tortoise-like in comparison with the first electronic computer, which appeared soon thereafter. This was the famous ENIAC (Electronic Numerical Integrator and Computer), designed and developed in the early to mid-1940s by John Mauchly and J. Presper Eckert Jr., both of the University of Pennsylvania's Moore School of Engineering (see plate 18). The ENIAC machine utilized vacuum tubes, which required no mechanical motions (such as with relays), and which performed the same multiplication several thousand times faster than the Harvard-IBM computer.

The ENIAC could perform so quickly because for the first time computations vacated the macrocosmic world and entered the microcosm. With macrocosmic electromechanical machines, the speed of calculations stays subject to the factors governing the friction and inertia of mechanical parts. An electromechanical relay, for instance, works by creating a small magnetic field, which then opens or closes a switch mechanically. These switches in turn activate other circuits. The closing of a relay switch requires anywhere from one to ten milliseconds (a millisecond is 1/1,000 second). In the microcosmic universe of the vacuum tube, the transistor, and later on the silicon chip, the only elements in motion are electrons, whose mass is so tiny that virtually no friction or inertia impedes their flow. For practical purposes their movement is nearly instantaneous. Accordingly, an electronic circuit is vastly more rapid than circuits in an electrical or electromechanical machine. With the ENIAC, a switching took only about five microseconds (a microsecond is 1/1,000,000—one-millionth of a second), at least one thousand times faster than the Harvard-IBM machine. Today these times are standardly measured in nanoseconds—billionths of a second—and beyond, leaving well behind the grandfather clock of our sensory-bound perception of time.

Within electronic digital computers, the symbols 1 and 0 (binary digits, or bits) encapsulate the microcosmic world of rapidly moving information. These symbols are expressed in circuits whose design derives from Boole's algebra and its central operations of logical multiplication, addition, and complementation: the logical 'and', 'or', and 'not'. In the last chapter we hinted at the possibilities of mapping 1 and 0 onto various classes of operational functions, such as the truth-value functions forming the propositional calculus of symbolic logic. The link between Boole and the operations of electrical (and electronic) circuits was the discovery of Claude E. Shannon (b. 1916), who later became the founder of information theory, a branch of electrical communications science. In 1938 as an MIT graduate student, Shannon published a paper based on his master's thesis in which he demonstrated that circuits could be analyzed by means of a two-valued Boolean algebra. Although others before him had formulated similar ideas, Shannon's paper became the catalyst for the development of "digital logical circuitry."

In digital circuitry, logical operations are achieved electronically by means of gates, or switches, which open and close, thereby halting or permitting the flow of electrons through them. The positions of the switches, either off or on, can correspond to a 1 or a 0, according to the design's architect. Consequently, placing a number of switches in a properly designed circuit enables the flow of current to correspond exactly to Boole's logic. For example, if we arrange them in a series, such as illustrated in figure 8.1.1, we can produce the 'and' of logical conjunction. In this case, current enters from the left along line L and passes through to the lamp (or any other apparatus) only when both switches A and B are closed. There are four possibilities in this pattern (A may be open or closed; B may be open or closed), and they match exactly the truth-value possibilities of logical conjunction in the propositional calculus, which we showed in the last chapter. That same table for conjunction, we also recall, maps directly onto the Boolean lattice of logical multiplication (or intersection).

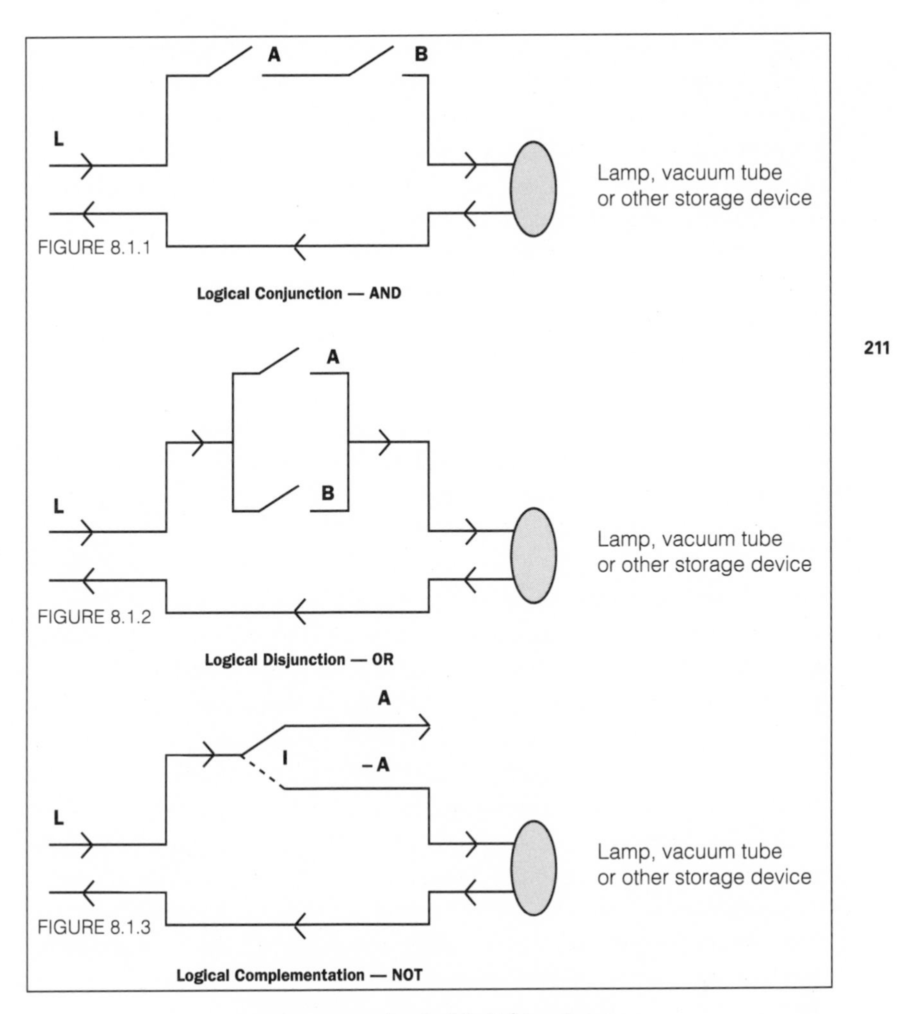

FIGURE 8.1. Logical Switching Gates

Just as placing switches "in series" achieves the correlation between electrical circuitry and logical conjunction, arranging them "in parallel" serves to map the operation of logical disjunction. In this case, shown in figure 8.1.2, current enters from the left, as with the previous example, only now it has two switches through which it might pass, two options for proceeding to its appointed end. If either switch A or switch B is closed, the circuit is completed. The resulting array of options corresponds exactly to the disjunction truth table of the propositional calculus and to Boole's logical addition.

In his pathbreaking paper, Shannon expanded much further the correlations between logical operations and circuitry, and demonstrated how relations like 'not' (negation), 'if . . . then', 'nand' (not both), 'equivalence', and many others could be cast in circuits. Thus, quickly, logical negation or complementation (figure 8.1.3) is achieved by a simple "invertor" (I) that negates or inverts the status of the current, turning it into its opposite, from on to off and vice versa. The result of Shannon's work was an extensive isomorphism between the propositional calculus and relay and switching circuits, an extremely effective design tool for computer hardware, as wiring circuitry has since come to be called. With it one could work out logical operations and combinations in the symbols of Boolean algebra and then translate them directly into electrical circuits. Such techniques were utilized in the early electromechanical devices and in the first generation of electronic machines, those using vacuum tubes. They still reside at the heart of integrated circuitry on the silicon chips of contemporary computers (see plate 19).

In the nineteenth century, the great (and often incomprehensible) German philosopher Hegel believed that changes in quantity eventually transform quality, that enough change in degree produces change in kind. Certainly such has been the case with digital speed. The advances in technology from Babbage's purely mechanical, analytical engine to the electromechanical devices of the 1940s to the electronic computers of our own day have augmented the speed of digital computing almost literally by quantum leaps. Computer scientist René Moreau, of IBM France, remarks that by the 1980s some IBM computers were functioning both over a thousand times faster and a thousand times cheaper than their Jurassic forebears of the early 1950s, with the combined improvement in relative speed and price totaling a factor of well over a million. Nor does a practical end to increasing the speed and volume of information "transfers" appear in sight (see plates 20 and 21).

This extraordinary speed of moving bits of symbolized information at the microcosmic level within the bowels of the computer has transformed qualitatively our understanding of information in both the micro- and macrocosmic realms. It has enabled the digital computer to become an all-purpose machine, extending its capabilities of storing and manipulating information bits far beyond those required for mathematical calculations. In turn the resulting sheer volume and variability of information transfers at the microcosmic, electronic level have created an information technology of hitherto unimagined capacity for registering, encoding, transmitting, and controlling information at the macrocosmic level of our linguistic capabilities and sensory imagination, the level of Aristotle's words and things.

Notwithstanding the enormous amount of commentary generated by the rapid growth of computer technology, we are just gradually beginning to appreciate seriously many of the profound intellectual and cultural dimensions of this transformation in the speed of information handling. Many of these implications grow directly out of the logic of micro-information, the basic syntactical rules governing the behavior of billions upon billions of electronic impulses. We must now turn to this logic and to its designers, Alan Turing and John von Neumann. The men were polar opposites in character and life stories, but the world of 1s and 0s cared not a fig about such matters. From their respective wellsprings of genius, both men (who did in fact know one another) shaped the procrustean logic and technology of the age. In their accomplishments reside the power and determinism of contemporary information.

Turing and the Logic of Micro-information

Throughout Alan Turing's abbreviated life (he committed suicide at age forty-two) he remained as much an enigma as the famous German cipher machine to whose decoding he devoted his wartime years. His biographer, Andrew Hodges, records that Turing considered himself something of a "heretic scientist, gloriously detached from the conventions of society in his quest for truth." The child of a declining British Empire, Turing's odyssey was indeed marked by imperial detachment and intensity. In his student years both found their most congenial supports in the rarefied, intellectual atmosphere of King's College at Cambridge University, that "lovely backwater" of the 1930s where Turing nurtured his penchant for mathematics, as well as his insouciance regarding worldly affairs. The wartime bunker of Bletchley Park offered him another hothouse retreat, a manor (rather comfortable at that) some fifty miles north of London where civil servants of the "professor type" spent the war secretly and frantically deciphering German coded messages. Both these havens provided Turing the free rein of his "terrific concentration."

Even before World War II, while Turing was at Cambridge that concentration had already been directed toward the topic of algorithms and computing machines. We have used the term 'algorithm' somewhat loosely and informally throughout our essay to designate any sequence of procedures or operations performed with symbols. Our usage reflects the varied history of such procedures. But in the 1930s Turing gave the concept of a general algorithm its precise formulation. Although several other descriptions were also proffered, Turing's conception of a universal computing machine has become synonymous with algorithm itself. The eponymous Turing machine refers both to the conceptual possibility of a universal, digital computer and to the systematic procedures of the central processing unit that operate the machine.

Turing conceived his machine to solve a complicated problem known to mathematicians as the halting or decision problem. In its nub the issue concerned whether there might exist a general algorithmic procedure for solving all mathematical questions in principle, or more accurately, for deciding in advance whether they could be solved. With his machine Turing proved such an algorithm could not exist. He demonstrated an unsolvable, undecidable problem, a "noncomputable" number, effectively establishing that no mechanical process of thought could determine in advance whether a computation, itself a mechanical process of thought, would ever end. Yet as Turing soon came to see, his universal machine (which was itself an exercise in pure thought) did exceedingly more than put to bed an abstruse problem in mathematical logic. It helped open a new era in thinking about information.

At the core of that thinking lay Turing's conception of a general algorithm as a purely "mechanical set of rules," which governed an "automatic" computing machine. Though it did not require human intervention, he believed the machine did have to resemble human thought in two basic ways, for "the behavior of the computer [understood here as a person] at any moment is determined by the symbols which he is observing, and his 'state of mind' at that moment." In Turing's idealization the machine's symbols were situated discretely, one by one, in a linear sequence of squares that comprised an infinitely long tape. Each square on the tape either contained a single mark or was blank (which we can depict as a 1 or a 0). The tape as a whole could thus store an unlimited amount of digitized information, arrayed in strings of binary symbols. (At this juncture Turing's proposal idealizes the matter, envisioning what is logically possible; eventually, electronic storage of memory bits would approximate the idealized, "infinite" tape to a remarkable degree.)

The machine's "state of mind" consisted of the operations it could perform on the tape. Their number was finite, although the actual number of symbols could extend indefinitely. Turing further conceived the machine's operations as "simple" ones, "so elementary that it is not easy to imagine them further divided." Like the symbols on the tape, the operations were also discrete, to be performed one at a time. They included first of all the ability to observe or "scan" (Turing's term) the squares so as to determine whether a 1 or a 0 were present in any given one of them. Next, if so ordered, the scanner could write over any symbol, changing it from 1 to 0 or from 0 to 1. Then it could move one square to either side. Continuing its movement one square at a time the scanner could relocate "to another square within L squares" of its previous location, settling on the specific site (or "address") of an information bit. In short, it could seek out new information at any given square on the tape. Finally, it could stop.

Turing subsequently explained these internal states, which he initially dubbed "states of mind," as governed by instructions: "The state of progress of the computation at any stage is completely determined by the note of instructions and the symbols on the tape." These instructions directed the endless moving, altering, combining, and recombining of symbolized information, arrayed in sequences of individual bits. They computed. With them one needed only sufficient time to process the bits discretely until reaching the end of a designated computation, after which the machine, as per the final command, would halt.

Discrete instructions and discrete symbols, then, comprise the mind and body of a universal Turing machine. A powerful and robust idea united them, namely, that there could be "no essential distinction between numbers and operations on numbers" (or between the instructions and the symbols—in current parlance, between the programs and the data). This idea signaled a consequence of momentous proportion, for it meant that the same abstract symbols of 1 and 0 could now express three distinct, but intertwined, features of micro-information processing: (1) the binary number system, which enabled the mechanical computer to calculate a limitless array of mathematical problems (given sufficient speed and volume of discrete operations); (2) the Boolean system of logical classification and truth-value functions, which embraced natural language and its subsumption under symbolic analysis; (3) the sequence of instructions, which provided the machine its marching orders, telling it what steps to take, what order to take them in, and when to stop. In brief, Turing's conception of a universal machine effectively synthesized the central techniques of processing the symbolized information of both numeracy and literacy.

To accomplish all this with only an infinite tape of 1 and 0 symbols and a scanning-writing device called for the development of elaborate and oftentimes complicated methods of coding. Turing himself had acquired considerable, hands-on experience with decoding machines during the war, and in the years immediately afterward he became extensively involved in the practical design of England's first electronic computer, the ACE (Automatic Computing Engine). Such activity led him to stress the importance of keeping the computer hardware "as simple as possible" and of buttressing it with a large memory for storage. He even warned against devoting hardware space to adders or multipliers because all calculations could be effectively handled with instructions using only the primitive logical functions, 'and', 'or', and 'not'. If the mind and body of the machine were its operations and symbols, he believed, its soul resided in the logical control of the coded instructions and information, in what is now termed software programming.

Encoding and computing the information on Turing machine tape meant a whole new way of thinking about binary digits, not just as numbers but also as encryptable symbols, ciphers capable of harboring an array of hidden messages. Binary arithmetic, the number system based on the two digits 1 and 0, had long been an instrument in the mathematicians' toolbox, dating from the days of Leibniz. Yet binary coding evolved only gradually through practical trial and error during the 1940s and 1950s. As it did, newly devised encoding techniques such as "contraction" and "expansion" became the effective means for inscribing numbers, words, and machine instructions side by side in the same string of 1s and 0s. Although many encoding techniques are extremely elaborate and intricate, we can illustrate the basic idea behind them with a single example, the procedure of "expansion."

Expansion enables us to take a series of numbers in common, decimal notation, together with the operations (instructions) we wish to perform on them, and to translate them into the coded 1s and 0s of a Turing tape, or its contemporary storage counterpart. Part of this procedure is direct and intuitive, but another part of it involves a trick. (Coding or enciphering of whatever sort is like magic. We first see the apparently mysterious results of legerdemain, but once the trick is revealed the mystery fades. Of course we still marvel at cleverness.)

Let us begin with a series of numbers, selected at random and separated by commas, and two operations, say addition and multiplication:

[A] $5, 6, 2 \times 7, 3 + 4,$

First we convert the numbers directly from decimal into binary notation. By contrast to numbers in a decimal system, the columns of a binary number represent powers of 2 ($1, 2, 2^2, 2^3, \ldots$). Decimal 5 thus equals binary 101 ($2^2 + 0 + 1$), while decimal 14 equals binary 1110 ($2^3 + 2^2 + 2 + 0$). Translating the entire string gives us

[B] $101, 110, 10 \times 111, 11 + 100,$

Now we face a complication. We need to introduce more 1s and 0s into the series in such a way as to represent the commas and operational symbols without making a hopeless mess of the numbers already written in binary notation. To do this requires that we establish a code for the commas and symbols. Arbitrarily, we shall let the number 2 stand for a comma, 3 for multiplication, and 4 for addition. Substituting, this yields the following string:

[C] 101 2 110 2 10 3 111 2 11 4 100 2.

This gets us closer, but notice that we still have to convert this string into an unambiguous sequence of 1s and 0s. And for this we need some legerdemain, it would appear, by which to translate the 2, the 3, and the 4 into binary digits. Were we just to convert them directly it would be impossible to discern where a number leaves off and an instruction begins. The whole string would become one gigantic binary number.

Our trick sounds complicated and magical, but it actually functions rather guilelessly to map the digits from string C onto a longer string. The principle is this: We make each individual digit in C correspond to the number of 1s lying between successive 0s in the expanded string D. (Note: Where a 0 exists in string C, this means there are no 1s between successive 0s in string D.) For convenience in visualizing the procedure, we have added spaces between the digits in string C:

[C] 1 0 1 2 1 1 0 2 1 0 3 1 1 1 2 1 1 4 1 0 0 2

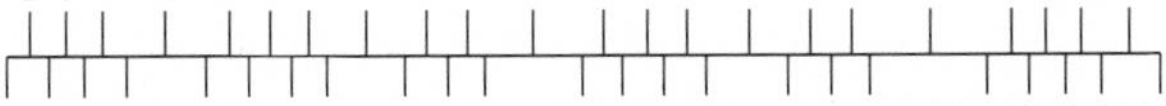

[D] 010 010 11 01010 0 11 010 0 111 0101010 11 01010 1111 010 0 0110

Each digit of the shorter string C represents the number of 1s lying between the 0s in the expanded string D. Finally, we remove the spaces from D and have our intended result, an exact translation of the original numbers and instructions into an optically daunting, but otherwise quite intelligible and unambiguous, sequence of 1s and 0s:

[E] 0100101101010011010011101010101101010111101000110.

This string of symbols now represents a "binary coding of numerical data" and the operations to be performed.

With the foregoing procedure we have shown how to encode into a string of 1s and 0s both numbers and the instructions for moving them about in various combinations. With enough space, speed, and time, and using the discrete instructions that Turing had identified, we can now calculate virtually any computable problem or function in mathematics. (Not all problems; recall that some are not computable.) The technique of contraction reverses this expansion procedure. It enables us to begin with a string of 1s and 0s coded with instructions and numbers, and to translate them into decimal notation and commonplace, operational symbols.

Enciphering a string of binary symbols allows a Turing machine to perform mathematical calculations automatically. But this serves as a bare introduction to the capabilities of a universal computer. We begin to broaden the horizons by looking at the foregoing in a slightly different fashion. Expansion and contraction define only two specific encoding procedures. They rest on the idea that in any given sequence of 1s and 0s, we can introduce coded instructions into the digits simply by placing two or more 1s in succession according to some rule or pattern. From this we may generalize and stipulate that different groups and patterns of 1s and 0s can be devised to perform many other, nonmathematical tasks. As cryptographers discovered centuries ago, numbers and number sequences can easily be made to correlate with letters of the alphabet, and therefore with words and messages. Using strings of binary digits facilitates the same process. They allow us to encode letters and the instructions for combining them into words and sentences, and eventually to create digital word processing. It becomes a relatively short step to add further instructions in order to create and manipulate abstract classes in accordance with Boole's logical algebra.

Turing's conception of a general-purpose computer has infiltrated our contemporary age as the core logic subtending micro-informational processing: the encoding and algorithmic manipulating of binary, digitized strings of information. From its birth in a human mind, this idealized invention with its "mechanical set of rules" has defined a new information technology, one capable of moving about a limitless amount of data. Since those heady days Turing's abstract machine has descended from its idealized, rarefied inception and steadily worked its way into the quotidian activities of our practical lives. The descent from ideal to concrete commanded the attention of another, equally Promethean figure concerned with computers and information, John von Neumann.

The von Neumann Flow

The popular image of mathematical genius depicts a mumbler, shy and often sockless, understood by few, an absent-minded eccentric who generally ends normal conversations by entering them. Witty, worldly, and jocular, articulate in seven languages and hypermnesic, von Neumann belied the caricature. Throughout his life, this Hungarian-born Mozart of mathematics delighted and flourished in the highest circles of intellect and society, as well as in the corridors of political and corporate power. In his life's journey from Budapest to Berlin to Princeton (where in 1933 he was among the first faculty appointments to the Institute for Advanced Study) to Washington D.C., he contributed an almost superhuman scientific talent to an enormous array of practical and theoretical projects. (A joke circulated for a while, in the Princeton of Einstein and Gödel no less, that von Neumann was not really a human being, but a demigod who had descended to earth, studied humans, and developed the talent for imitating them perfectly.) Scores of von Neumann tales revolve around his wizardry, especially his photographic memory and uncanny calculating ability, and over time they have assumed nearly legendary proportion. Most are probably true. Awe and anecdotes aside, the words of von Neumann's biographer probably best describe him, simply, as a "paragon of science."

The necessities mothering, or at least midwifing, the birth of the electronic computer and of contemporary information derived from science, engineering, and business, but in the end chiefly from war. Like Turing, von Neumann's concentrated involvement with computers began during World War II when he and many other scientists, engineers, and ballistics specialists needed to process huge volumes of calculations. In his case the mathematical work related in particular to the triggering mechanism of the atomic bomb. But just as the Turing machine quickly superseded the decision problem that had led to its invention, so too did von Neumann's accomplishments in computer design quickly move beyond their wartime beginnings. The logic of Turing's idealized computer had provided the conceptual blueprint of a universal information machine. The logic of von Neumann's machine structured the entire arrangement and organization of the component parts for making the computer effective and efficient.

Von Neumann first sketched these components in a 1945 working memorandum that summarized discussions among the Moore School engineers and scientists who had developed the ENIAC. (He had become a consultant to the group and was participating in the final phase of the ENIAC's construction.) The following year von Neumann expanded his sketch into a longer, technical report for the army, jointly authored with Arthur W. Burks and Herman H. Goldstine. These working papers outlined the design of a machine known as the EDVAC (Electronic Discrete Variable Computer), which was the immediate antecedent of the IAS (Institute for Advanced Study) Computer at Princeton. Fully exhibiting what is now standardly termed "von Neumann architecture," the latter stands as a prototype of today's systems. Both documents identified the key components of the modern computer: the central processing unit (the CPU, which includes arithmetic calculators and control devices), the memory storage, and the input and output mechanisms for bringing information into the CPU and reporting its results.

More important, in these and subsequent writings von Neumann detailed an overall plan for the logical and sequential operations of the computer. To this end he developed the "flow diagram," which charts the course of information transfers and operations among the machine's different parts. Central to the flow is the memory, technologically made up of "microscopic elements of some suitable organ," capable of storing digitized information electronically. Encoded binary information bits are introduced into the memory through some type of an input device, which serves to translate a computable problem into the language of the machine. This can be accomplished in many by-now-familiar ways with punch cards, typewriter keyboards, optical scanners, and so on. The bits include numbers and words, of course, but more significantly, the coded instructions that direct the operations of the central processing and calculating unit. Von Neumann called this critical feature of his design the "memory-stored control." A commonplace today, his arrangement thus stores both operating programs and information files together in the same memory.

With information once placed in the memory, designated portions of it (all still in binary code) are then transferred to the CPU, according to the instructions, where they are processed. Afterward, the processed results are returned to the memory and subsequently are sent back to the operator by means of an output device. The two most frequently used means for this last procedure now are the CRT, the cathode-ray tube of the computer monitor, and the printer for "hard copy," paper printouts. In this scheme, then, the memory sits between the human operator and the actual processing, and information flows in two directions: (1) from human operator to input to memory to CPU, and (2) from CPU back to memory to output to operator.

Given this overall plan one can easily see why the logical order of memory control commanded so much of von Neumann's attention. In effect computer "memory" has become synonymous with encryptable procedure, with algorithm. We still use the metaphors of storage and container (as did von Neumann), but these stand, anachronistically, as the projections of a literate mind into a universe of pure symbol manipulations. The actual workings of memory comprise the logical steps that process encodable experience—instructions, numbers, words—into the patterns of 1s and 0s, which are processed further in the CPU. In this way substance is transformed into logical rules, the core of the computer's power. Memory "storage," then, simply designates a critical locus in the flow of digitized information.

Within the memory itself, von Neumann assigned to each string of information bits, whether number or word or instruction, a specific "memory register" at a "definite address." From their initial step, the instructions discretely guide their own march through the registers as each one singularly determines its successor. This is accomplished in one of two ways, unconditionally or conditionally. Unconditional instructions simply carry out their assigned operation and then move to the next address in the sequence of registers, seriatim, continuing one by one until they have completed all their tasks and stopped. This pattern von Neumann called processing through "control sequence points."

By contrast, conditional instructions advance through "branching points," with each new branch and each new instruction depending on the specific outcome of the previous command. This instruction pattern produces a "logical tree," which also goes by the more familiar term "conditional branching." Conditional branching, the logical "if . . . then," provides the pulse beat of von Neumann's design. With it the sequence of instructions flows, not through a preordained, fixed series of registers, but directly from the results of each preceding instruction. Automatic machine "decisions" about information transfers, then, can vary from instruction to instruction according to the requirements of the overall task.

To show more concretely how conditional branching works, let us step for the moment outside the information microcosm of binary digits and follow a small computer program in ordinary language, which we can represent with a flow diagram greatly simplified from those of von Neumann. For this task we shall assume one of the economists' omnipresent widget companies, one with a large sales force of several hundred people. (Computer hawkeyes, please note: The example will be less than elegant, or even precise. We wish mainly to illustrate how information proceeds through programs, not how to design them.)

Into the computer's memory we place a record pertaining to each of the company's salespersons and containing the following information: sales territory (by state), amount of sales for a month, and gender. We could always supplement this "data entry" with additional items (commission structure, salary, years with the firm, customer complaints, whatever), but the above will suffice for the model. Each person's record resides in a specific register (or address) in the computer's memory. Collectively these records comprise a set with sequentially ordered members. In the flow chart of the example (see figure 8.2) we have designated the set as "Salespersons A, B, C, . . . N." Now we can devise a list of instructions for proceeding step by step through the records in order to discover different, perhaps less obvious information about the employees and their activities. The list of instructions will constitute the program's algorithm.

Suppose we wish to identify the female employee with the highest sales total in New York state for the month. We start the program by first introducing

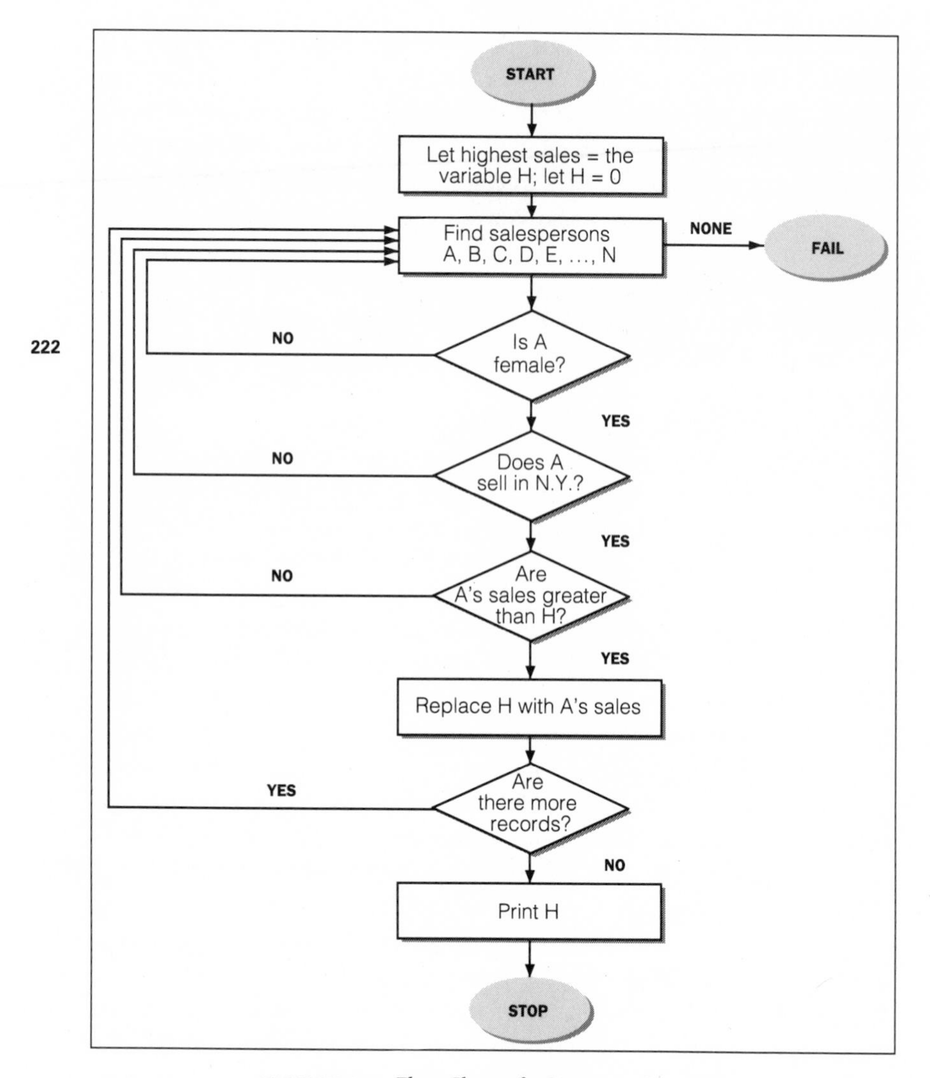

FIGURE 8.2. Flow Chart of a Computer Program

an instruction, which is to have the letter H serve as a variable dollar amount representing the highest sales, what we are trying to discover. Initially, before searching any records, we set this value at 0. Next we instruct the computer to find the first salesperson (salesperson A) in the memory. When it has done this, we give it a conditional instruction. It takes the form of a hypothetical, 'if . . . then' statement: if A is female, then proceed to the next instruction; if A is not female then return to the set of employee records and proceed with the next record in the sequence (in this case salesperson B). In the flow chart, we have written the conditional commands in question form and placed them in diamonds. Each question has two possible answers, yes or no, and each successive step in the program depends upon which answer is given.

The instructions and questions enable us to continue one step at a time until the completion of the overall assignment. We traverse the file of salespersons, selecting those who are female and who sell in New York. With each one we compare her sales total with the highest sales value up to that point and substitute each higher total as it is discovered. Finally, when the program has completed its sequential search through the records, it instructs the output device to print the resulting H and tells the program to stop. (Remember, all this happens within an eye twinkle; it generally takes much longer to explain how programs work than to run them.)

We have smuggled several assumptions into our rhetorical example. Precise coding in a computer language would eliminate them with further steps and details. Nevertheless, the example illustrates how von Neumann's memory-stored "control sequence points" and control "branching points" enable information to be moved about and processed. Also capable of driving mathematical calculations, these sequence and branching controls rest on the primitive, logical connectives of 'and', 'or', and 'not' that Boole had defined over a century ago.

Other dimensions of machine-level binary programming grow out of these two controls. There are, for instance, many variations of conditional branching. In one, the branches continue subdividing and retrieving different record files, rather than looping back to the same, initial record set. This is a valuable tool for searching various databases and creating new ones from them. In another (commonly used in chess programs and the like) branching leads to the "deeper" levels of a memory hierarchy, with each new level containing a geometric increase in the volume of stored information and operational choices. Von Neumann's sequential design, further, makes possible the use of "subroutines," processing "loops" that can perform side tasks tangential to the main program and then bring one back to it. (Perhaps we want the telephone number belonging to the highest saleswoman; we could introduce a subroutine to retrieve and print it along with the final result.) The ugly word 'interface' has even crept into our linguistic nest under pretext of describing more elaborate methods of conditional branching. These methods can unite entire programs, databases, and computing systems, whose only restrictions are available memory and speed of operation.

Since the early days of von Neumann's plans, the basic architecture of micro-information processing in sequential steps has stayed remarkably true to his schema. Despite the technology's enormous increases in speed and volume of processing with the advent of semiconductors and the integrated circuits of silicon chips, most contemporary computers represent some version of the machine he conceived in the 1940s. The development in the 1970s of the commonly used random-access memory chip (RAM) has changed the technology of how information bits are stored and retrieved, substituting random storage for fixed memory registry (see plate 22). But even so, the same rules of processing remain as binding as ever. (The greatest departures from von Neumann's design have been those exploring the avenues of processing digits simultaneously in parallel pathways and "neural networks.")

Much like Turing, von Neumann insisted on keeping the hardware reasonably simple and on devoting attention to the critical issues of memory size and software programming. In contemporary computer design, the logic of "memory organization" still outweighs in importance the material technology of the computer's construction. Computer specialists term the challenge of creating fast, efficient, and effective algorithms for operating systems and programs simply as the "software problem."

The Power of Pure Technique

In the foregoing we have shown how Turing and von Neumann developed the logic of micro-information, which permits us to encipher, arrange, store, and process strings of binary symbols. By itself digital information represents nothing more than the symbols 1 and 0, bits placed together in various sequences and patterns. We have referred to these symbols as "information" because they do indeed comprise the microcosmic information within the computer. Their more familiar appellation is "data," which identifies them as the smallest constituents of information that we can transfer about, or compute electronically. Binary, digitized data have become the new informational "given" of our contemporary culture.

Yet, as we suggested at the outset of this chapter, data occupy a very peculiar place in the information "hierarchy" (as it has been tagged by Bell Laboratory's Robert W. Lucky and others). This is because they are also empty, meaningless containers without contents. We are reminded of a famous passage in which Bertrand Russell once described an atom as consisting "entirely of the radiations which come out of it. It is useless to argue that radiations cannot come out of nothing. . . . The idea that there is a little hard lump there, which *is* the electron or proton, is an illegitimate intrusion of commonsense notions derived from touch. . . . 'Matter' is a convenient formula for describing what happens where it isn't."

An information datum mirrors the atom. Information comes out of it, as it were, once we begin processing it in combination with other data. At the microcosmic, digital level "information" stands as a "convenient formula for describing what happens where it isn't." Like the Cheshire cat, both an atom and a datum fade before us as we seek their hard lumps, only to reappear again as we turn our backs and start applying the rules of logic.

Logic binds digital data; it is the set of rules according to which the data symbols may be moved. Once in "motion" data comprise algorithms. And each step in an algorithm must be unambiguous and rigorous, logically necessary, following without exception from its predecessor to which it is chained by the rules. The steps of this movement comprise our information age's internal "chains of reason," its central power. Only through its exercise can there emerge the patterned strings of 1s and 0s by whose means we can encode the "stuff" of our exchanges with the world. At the deepest level, then, the logical, algorithmic power of computer technology actually constitutes information.

In contrast to the world of atomic physics, however, the macro-information that radiates from otherwise empty digits rests entirely upon the macro-information that has been poured into them. This requires a twofold operation of breaking down the phenomena of our sensory world into smaller and smaller components and of correlating these components with patterned sequences of 1s and 0s, of encoding digitally the words and things of our experience. In previous ages, scribes, copyists, clerks, printers, human computers, and other information technicians all possessed guidelines for closure in their own idioms. Yet, today's user of information technology finds no single, correct, classifying schema, no universal, mathematical "mapping" of a fixed, informational world that can serve as a guide through the contemporary morass that makes up our own information overload. There are only data, endless data to be encoded and shuffled about.

Indeed, mammoth, electronic databases now house libraries of information in digitized form, all of which can be retrieved with Boolean methods and operations that enable the user to create new encyclopedias virtually at will or whim. Whole new information "universes" can be constructed, connected, and controlled by users of the technology. Such developments have led some observers to fear the tyranny of information control exercised by an Orwellian Big Brother, like the FBI or the IRS or the credit snoop TRW, others to tout a political democratization of information for all, a "computopia" wherein voluntary, "self-actualization" communities will finally achieve a pollution-free "symbiosis" with nature. Yet fears and hopes aside (neither of which will probably pan out quite as we envision, since nature and humankind alike love the hidden flaw), the absence of fixed foundations for this new technology, beyond the logical and material techniques of micro-informational processing, opens up the second, macro-informational dimension of our age: the realm of play.

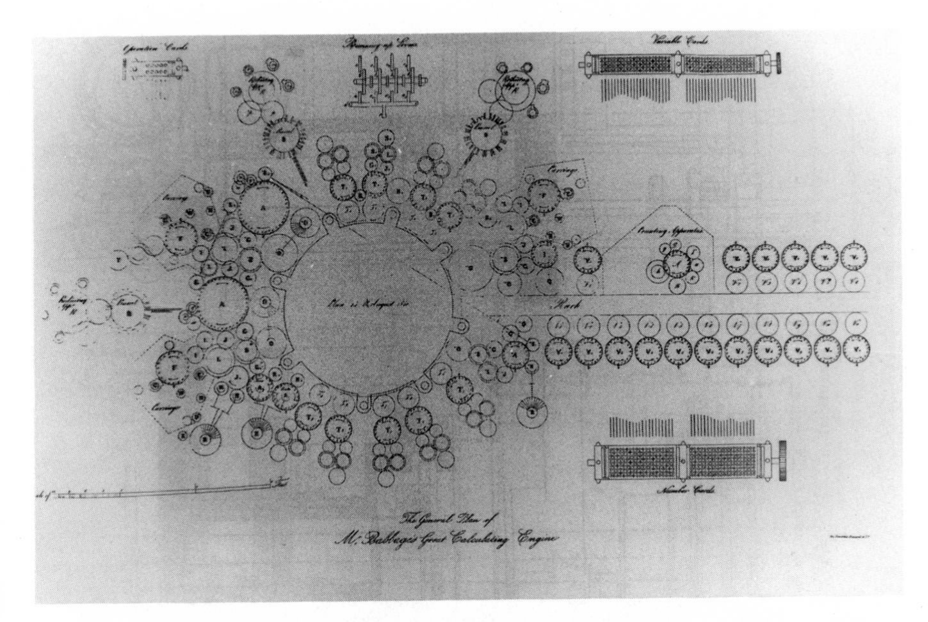

PLATE 16. *Babbage's Analytical Engine—General plan of Engine No. 1.* Dated August 6, 1840, "Plan 25" depicted, in effect, the first "digital" computer. The wheels surrounding the large circle on the left made up the "mill" (CPU), while the wheels on both sides of the rack constituted the "store" (memory) of the device. In theory, the rack and storage wheels could be extended indefinitely; in engineering practice, the amount of power and torque needed for the long, mechanical drive train remained a problem until the twentieth century. As the initial gears and wheels turned, power and torque quickly dissipated and became inadequate for impelling the remaining parts of the apparatus. From Philip Morrison and Emily Morrison, eds., *Charles Babbage, On the Principles and Development of the Calculator, and Other Seminal Writings by Charles Babbage and Others* (New York: Dover, 1961), pp. 378–79.

PLATE 17. *Babbage's Analytical Engine—the "Mill."* Based on Babbage's drawings and under the supervision of his son, this working model of the mill was completed not long after Babbage's death. From the Science Museum, London.

PLATE 18. *The ENIAC (Electronic Numerical Integrator and Computer).* Designed and developed in the early to mid-1940s by John Mauchly and J. Presper Eckert Jr. at the University of Pennsylvania's Moore School of Engineering, the ENIAC was housed in a room of 30 × 50 feet. The first working electronic computer, it consisted of nine basic units—forty panels with 18,000 vacuum tubes, 70,000 resisters, 10,000 capacitors, 6,000 switches, and 1,500 relays—and was powered by 150 kilowatts of electricity. It could perform a stunning five thousand basic arithmetic operations in a single second. Here are the ENIAC's cycling unit and accumulators (slightly less than half the overall machine). From the Smithsonian Institution.

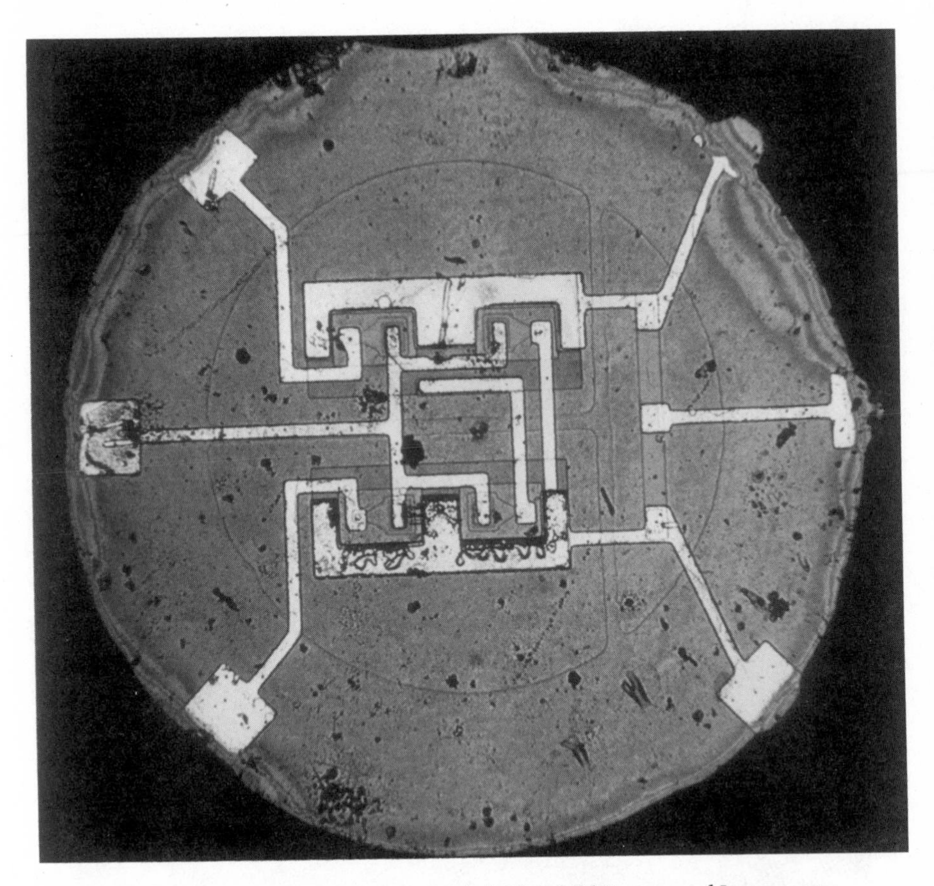

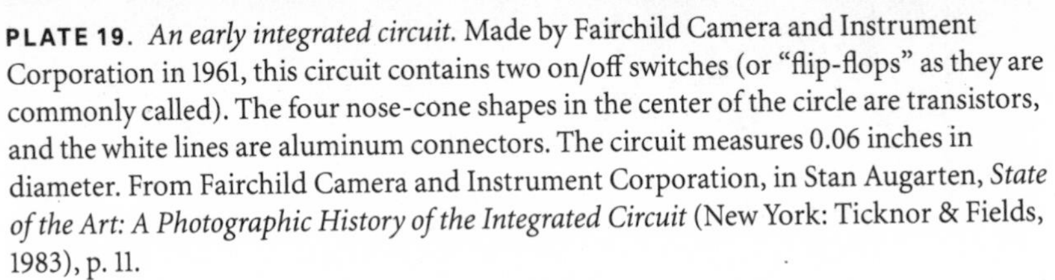

PLATE 19. *An early integrated circuit.* Made by Fairchild Camera and Instrument Corporation in 1961, this circuit contains two on/off switches (or "flip-flops" as they are commonly called). The four nose-cone shapes in the center of the circle are transistors, and the white lines are aluminum connectors. The circuit measures 0.06 inches in diameter. From Fairchild Camera and Instrument Corporation, in Stan Augarten, *State of the Art: A Photographic History of the Integrated Circuit* (New York: Ticknor & Fields, 1983), p. 11.

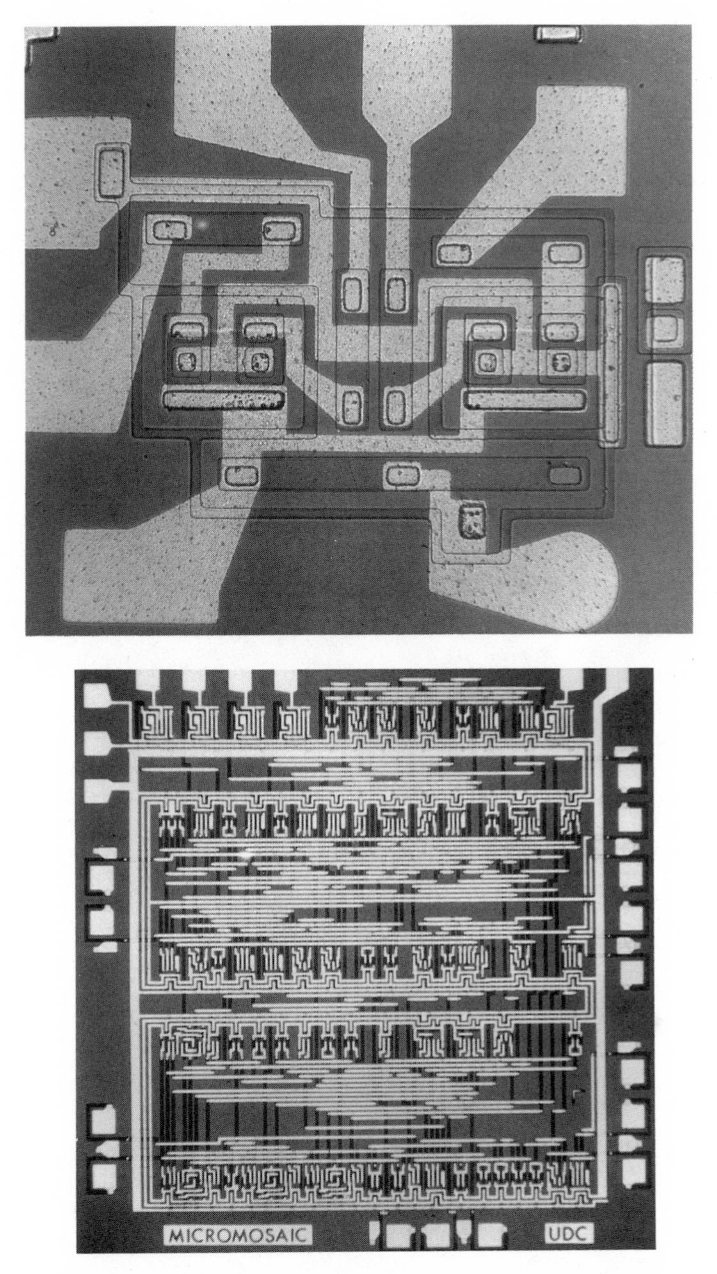

PLATE 20. *Early logic chips.* Both of these chips were developed and produced by Fairchild Camera. Appearing in 1963, the chip on top was rather simple, containing only four on/off switches. Four years later, its offspring was far more complicated and used a combination of transistors (the dark vertical lines) and aluminum interconnections (the light horizontal lines) to create over 150 logical gates in a space of 0.15 square inches. From Fairchild Camera and Instrument Corporation, in Augarten, *State of the Art,* pp. 15, 23.

PLATE 21. *Microprocessor.* Developed in 1970 by the electronics firm Intel, the "4004" microprocessor, which measures approximately 0.1 inch wide by 0.15 inch long, was the first "computer on a chip." In 1977 its more sophisticated heir, the "8080," became the first chip powerful enough to drive a personal computer. From the Intel Corporation, in Martin Campbell-Kelly and William Aspray, *Computer: A History of the Information Machine* (New York: Basic Books, 1996), center photo gallery, between pp. 150 and 151.

PLATE 22. *RAM (Random Access Memory).* Where is the information? A blow-up of a section from an IBM high-speed, 64K RAM chip. This chip can process a sixteen-bit word in between 15 and 20 nanoseconds (billionths of a second). The section here serves to detect and amplify the coded electrons stored in the chip's memory. Corresponding to 1s and 0s, the electrons carry instructions and data to the central processing unit. From IBM, in Augarten, *Bit by Bit,* p. 282.

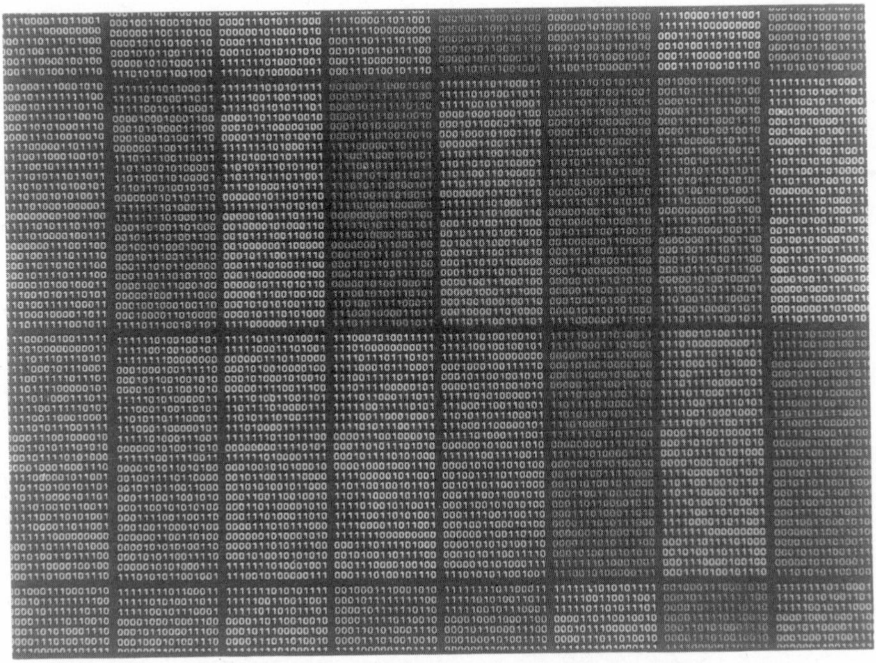

PLATE 23. *Digital code.* In the contemporary age, information is any phenomena, including sound, that can be captured in digital code. Here, for example, are the first four notes of Beethoven's Fifth Symphony encoded as a string of 1s and 0s. From the 3M Company, in Augarten, *Bit by Bit,* p. 35.

CHAPTER NINE

Information Play

All play has its rules. —JOHAN HUIZINGA, *Homo Ludens*

No one today can predict what games post-Gutenberg man will want to play. —GORE VIDAL, *United States*

From Power to Play

With their digital strings laced tightly in the Spanish boot of computer logic, today's data reveal the power of contemporary information technology. When extended to the human-scale realm of macro-information, the kingdom of our senses and imagination, the same data reveal power's complement: information play. Computer power gives us the black-white, on-off, distilled and abstract starkness of our information age, while play bursts forth in an extravaganza of color and hue, painting a richly textured information canvas.

It may appear odd to assert play, along with power, as the essential features of an information technology, comparable to the wisdom and knowledge of earlier ages. Play in particular seems a weak counterpart, largely because we tend to associate it either with frivolity or, maybe more seriously, with the theater. Yet, play exhibits many dimensions that supersede these two, most common of its guises. Recall the noted Dutch historian of culture Johan Huizinga, who depicted play as "voluntary activity," which we as *homo ludens*—man the player—engage in according "to rules freely accepted but absolutely binding," an activity carrying the "consciousness that it is 'different' from 'ordinary life.'"

Huizinga's incisive remarks point to a deeper, intellectual and cultural significance of play as an activity permeating today's information technology.

Computers are enormously powerful information-processing machines; they are also, culturally speaking, toys. (No surprise: computer games and amusements still command the lion's share of the technology's mass-market sales.) When we "play" with them we follow freely the "absolutely binding," logical rules of information processing, which epitomizes play activity. Moreover, as play, this rule-governed activity justifies itself; it needs no further rationale. We engage in it simply because "it is there," and because we have the ability to do so. Therein, at least in part, dwells the perception of fun as the psychological companion of computer play: the captivation, the tension, the enjoyment, and the awareness of play's difference from "ordinary life" that Huizinga observed.

Yet, more than psychology is at play in play. Among its many senses, play connotes looseness. A loose nut in a machine has play; a nut on the loose plays havoc with social order; a bad pun is a play on the loose meanings of words. Writ large, such looseness in today's information derives from its lack of closure in knowledge, from the arbitrariness of symbols, and from the union of digital symbol with electronic computer, a union whose high-speed and algorithmic, iterating capabilities can generate completely novel forms of information.

This looseness spreads throughout three, more focused manifestations of play that define our age: profusion, complexity, and emergence. Each of these terms mines a progressively deeper and richer vein of intellectual and cultural play that issues from the computer's micro-informational power and determinism. In digitizing macro-information, the exercise of raw computer power results directly in these extensions. Along with power, then, play constitutes information. Together they have completed the severing of information from its centuries-old, intuitive meaning as the shaped or informed kernel of content we elicit from experience and reconnect with the world outside. Now, with their logical rules and electronic speed, power and play simply bracket our exchanges with the world.

Perhaps the most immediately striking feature of today's information is its unprecedented profusion, which stems directly from computer technology. Even in our era of such rapid changes, no other technological innovation (except, possibly, television) has come so far so fast to intrude so much in daily affairs. Political, economic, social, and cultural imperatives, of course, have all taken a turn at giving rise to this profusion, but play has been its most constant impetus. This is the creative, magical play of human curiosity and exploration, of testing and logic, of imagination and invention, of discovery and application—all directed toward digitizing human experience. Converting matter to electronic signals, "atoms" to "bits" (in the words of Nicholas Negroponte, founder of MIT's Media Lab), we engross ourselves in these activities simply because we can. This play has radically transformed information technology and placed before us a cornucopia (or cyclone) of information beyond compare.

Although confronting us with the most abundant evidence of play, profusion only reveals play in its lowest form. The activity of play itself highlights a second, deeper dimension: complexity. Complexity puts today's information age into sharper relief against the backdrop of its predecessors. At the core of the classical and modern ages lie two contrasting idioms of organized information: wisdom and knowledge. Grounded in either alphabetic literacy and classification, or in numeracy and analytical abstraction, these idioms were generally believed to define information structures, timeless and true, notwithstanding their capacity to capture change as well as permanence.

An opposing idiom of information inhabits the contemporary age. It appears not as a fixed structure, timeless and true, but rather as ongoing, playful activity, time- and place-bound, teeming with local rather than universal truths. In today's science, this activity can be seen in the mathematical innovations associated with "chaos" and in the closely related sciences of "complexity." These new sciences rely heavily on iterative, computer simulations for their modeling of natural phenomena. They often begin with simple algorithms or rules or formulas, then repeat them at great speeds and to great lengths. This computer play yields not the linear, "predictable" end product of formulas and equations in the analytical mode, but new patterns, unpredictable behaviors of complex phenomena, and random, chaotic calculations and events. Even an omniscient God could not know the exact outcome of such repeated dice rolls.

Although broadly interpreted, complexity generally involves assigning quantitative measures to phenomena that fall somewhere between organized "steady states" and random "chaos." Yet studies in complexity often lead beyond merely quantitative measures into a third and even richer dimension of today's information play: emergence. Computer technology has now become a powerful tool for unearthing novel, unpredictable, and emergent patterns and structures of phenomena. These go by various, related names, such as "self-organization" and "complex, adaptive systems," and they share a common property. The very appearance of such natural systems results from their continual processing of ever-changing information, which returns to them through "feedback loops" as they adjust and adapt to their surroundings. (In fact, many scientists also propose analyzing developments in the human world as similarly emergent phenomena.) The actual discovery of emergent systems springs from computer simulations whose algorithms process results fed back from previous computations. With emergence, information play compounds itself in the service of new discoveries.

Emergent systems share a further property: They follow the "arrow of time," the irreversibility of the natural and social processes of which we are a part. In our essay, we too have followed the arrow of time, albeit loosely, from the dawn of information to now. Commenting on one's own age generally proves to be tricky business, all the more so when the topics involve dizzying, technological change. For our final chapter, we have selected items that appear to stand out historically in relief from the welter of available material, and we believe these will have some staying power so as to be acknowledged at a later moment. Yet we are students of history, not prognosticators. We venture no clairvoyance about third waves and megatrends. We are content to round out our gaze upon the past with its culmination in the present and to end with the unpredictability of emergence itself. For despite the virtually unimaginable volume of existing information we face today, our contemporary information age has just begun.

The Profusion of Play

The sheer profusion of computer technology obviates our having to describe it in great detail. Its "formless data" of "bits," enthuses Negroponte, have arrived from inner space and are everywhere here to stay. Unless one is a fin-de-siècle Rip van Winkle, awakening after a hiatus in consciousness for the past twenty years, we all "know about" computers, information, and "being digital." Whether surfing the Internet, paying VISA bills, getting cash from an ATM, connecting a phone, buying airline tickets, making book, playing Nintendo, or what have you, our ubiquitous computer technology inundates us with information and directs our lives, with or without our "input."

Although the omnipresence of computer products affects us most in our daily affairs, it discloses only the most basic level of information play. Huizinga's penetrating comment that "culture arises in the form of play" applies appositely to our computer age. Information profusion exposes the "form of play" that we engage in for no reason other than its possibility. This play is whimsical (as opposed to frivolous), even though we may engage it with utmost seriousness, for solemn purposes, or from weighty motivations. It is there for us to do, and we do it. Once having encoded and processed a slice of experience into digital strings, we can then forget about it and manage it by rote, moving on to the next slice. Pace by petty pace, imperiously and horizontally as it were, we extend the computer's digitizing power over more and more segments of macro-information. Our culture's digitally processed information thus pours forth and with its profusion of products "proceeds in the shape and the mood of play."

Two observations help us place this form of information play into some perspective. The first concerns the means whereby bits connect to the human-scale world of macro-information, the world of our senses and imagination, and the second concerns what is new and old in this information technology. Ever since the early days of computers, the task of connecting the micro- and macro-informational worlds has fallen steadily to special, "programming languages," which have been created to govern the flow of macro-information into and out of the computer's electronic interior. Encoding experience into digital strings, computer languages operate at two different levels, "low" and "high" (labeled thus with a singular lack of imagination). "Low-level" or machine languages function immediately within the universe of coded binary digits, which map directly onto the electronic circuitry of the computer, or whatever other processing organs there might be. These were particularly important during the infant years of computer development as designers translated directly into coded 1s and 0s the numbers, words, and instructions of the information to be processed.

Out of these formative efforts emerged "assembly languages," which served in the 1950s as the first programs to translate information between human operators and digital, machine language. Such innovations utilized systematic, mnemonic codes that eased conversion of problems into digitized form. For example, *S*(*n*)*C*, might mean 'store in the *n*th register of the memory an information content *C*'. The *S*, *n*, and *C*, would then each be represented by a specific pattern of 1s and 0s that was included in every string of coded digits. A critical deficiency restricted this procedure, though, because assembly languages remained dependent on the specific machines for which they were designed.

Along with greater standardization and sophistication of computer hardware, "high-level" compiler languages were first developed in the late 1950s and '60s. These carried, and still carry, an enormous advantage over earlier assembly and machine languages, because they are "machine-independent." Programmers can concentrate on working directly in the programming language and use the resultant, specific programs on different computers. Software design of new information programs can then develop independently of actual machine construction. The widely adopted FORTRAN (short for Formula Translator) and COBOL (Common Business Oriented Language) were among the earliest of such high-level languages. Used for the direct processing of scientific and commercial information, respectively, these stately flagships led a flotilla of high-level languages that followed in their wake.

Combined with technological advances in computer hardware, the liberation of programming languages from their early machine dependence has brought about the profusion of macro-information we find today. We can only hint at the technical virtuosity now available in these languages and programming, and at their role in converting macro-information into bits for processing. Indeed, we have been forced to bypass entirely many relevant and interesting topics, such as the digital encoding and processing of audio and visual information, developments that have sired whole industries (see plate 23). Even as we concentrate on computer language, the rapidly changing technology of programming and machine design threatens our commentary with the scourge of obsolescence.

A historical perspective lets us stay the course, however, at least a while. For the central purpose governing all computer languages—low or high level, specific or general, user-friendly or hair-pullingly convoluted—continues to be the effective translation of macro-informational content into the digitized world of micro-informational processing, and the reconversion of the latter's results into a form useful for the computer's operator. At the macro-level, in sum, computer languages provide the tools of play in contemporary information technology. Through them our experience becomes digitally encoded and processed in an ever-expanding profusion of information.

Our second means of placing the profusion of information play into some perspective focuses on what is new and what is old in today's ubiquitous information technology. Its applications have resulted both in fresh ways of expressing old information and in new forms of information itself, which display countless permutations. Computers now store and process an endless variety of records and information formerly managed by hand. But much of this informational content has altered very little. Businesses still process payrolls and invoices; banks still loan money at higher interest rates than they return to their depositors; factories still keep tabs on raw materials and products; hospitals still store and retrieve medical records and file insurance claims; schools still pass out grades. A C+ remains a C+, whether calculated with a short pencil or a computer grading program. Nor do computers actually manage macro-information in ways dissimilar from earlier, more cumbersome methods. In the words of Nobel Prize winner Arno Penzias, computers mostly perform "existing jobs more effectively and conveniently, rather than in a fundamentally different way."

Even while facilitating old forms of information management, computer play creates new kinds of information. Hegel's observation that at some point quantity will effect quality surely applies here; faster *is* different. Examples and images proliferate. Think only of spreadsheets in business forecasting, without which the calculations needed for leveraged buyouts or similar transactions would quickly overwhelm an office staff. Or of writing with a word processor and its ease of storing, erasing, combining, moving, exchanging, and traversing texts and "hypertexts" in "cyberspace." Or of expert systems that, with their branching logic, diagnose diseases and purchase stock options. Or of searching mammoth databases, such as developed by the genome project, for previously unknown connections between "raw" data. Or of simulated, virtual realities. Or of computer graphics, desktop publishing, and budget analysis. Or . . . one can supplement the list as desired.

In these and like cases, the lightning speed and massive volume of information processing dramatically transform the nature of macro-information in countless ways. From existing bodies of data one can generate new facts, patterns, predictions, and connections, all of which would be otherwise indiscernible or unavailable. The new information forms vary infinitely, and their specific uses depend entirely on the creators and managers of the programs themselves (giving rise to a twist on Disraeli's three kinds of lies: lies, damned lies, and information). The tasks involved in new information discovery may be and often are serious, their processing the result of information play, whose future appears boundless.

Chaos and the Sciences of Complexity

With the explosion of current information technology, the idea of information itself has metamorphosed into a new dominant metaphor or driving image or master idea (what the French term an *idée force*) behind much of our intellectual and cultural life. In the view of many, the metaphor has come to challenge, to equal, or even to replace that of scientific law. From the early scientific revolution of the seventeenth century through the great age of scientific determinism in the nineteenth, the successful discovery of patterned, predictable "laws" in nature had lured scientists, whose achievements loomed as the pinnacle (critics said the "Lorelei") of human knowledge. As metaphor, law anthropomorphized phenomena by extending terms associated with human and social behavior to the universe of impersonal events. "Nature to be commanded must be obeyed," wrote Francis Bacon some four centuries ago. But neither he nor other early scientists and natural philosophers really believed that things consciously "obeyed" the "laws" of nature in the sense of willful compliance to writ or command. Rather, the expression typified a human way of speaking about patterned, natural phenomena, one that brought events under human purview and, frequently, control.

Much like scientific law, the metaphor of information and its processing resists precise, general definition, even while lending itself to a myriad of very specific expressions and exact calculations. From its birth in computer technology, it too has begun with the human world of agents and activities and now reaches far beyond into the completely impersonal realm of natural phenomena. Physicist Murray Gell-Mann, Nobel laureate and quark discoverer ("he has five brains and each one is smarter than yours"), even suggests that one might look at the entire universe "from the point of view of information." This capacious vista unites the reductionist, fundamental laws of physics with accounts of emerging and "complex adaptive systems." In the life sciences, comparably, some biologists now see information processing as the most basic explanation for the behavior of phenomena ranging from biochemical molecules to cell and organ formation to the complex structures of entire organisms and their interactions with the environment.

The elevation of 'information' from a descriptive term into a dominant, explanatory metaphor has invested information play with a broader and deeper significance than that of its sheer profusion. This play may be found in the new mathematics of chaos and in the new sciences of complexity. (Gell-Mann himself cofounded the Santa Fe Institute, which is devoted to the study of complexity.) New fields and topics, new research programs and strategies, new images and terminology all attest the burgeoning sciences: turbulence and nonlinear dynamics; autocatalysis and self-organization; nonequilibrium, open thermodynamic systems; phase transitions into complexity; the butterfly effect and life at the edge of chaos; fractals, strange attractors, and much more. In contrast to the "classical" sciences, these approaches tend to be synthetic and holistic rather that reductionist, complex rather than linear, and unpredictably emergent rather than predictably determined. As lay scientists, such topics offer us convincing and fascinating evidence of something extraordinary occurring. Clearly, Watson, the game is afoot.

In nearing the end of our historical essay, we need to situate these scientific developments firmly within our own information age, just as we did other idioms of science in the classical and modern ages. Driving the new sciences of complexity is the all-important practice of information play, which flows directly out of computational power. Here play entails high-speed iterations and the results elicited from devising a computer program, or algorithm, and then turning it loose, as it were, to see what ensues. The wile of these playful algorithms springs from carrying out an initial computation and then feeding its results back into the algorithm in successive performances of the same computation. The "output" of the first cycle becomes the "input" of the next go around, and on and on. In mathematical terms, some algorithms behaving this way fall into a class known as "finite difference equations." (Nonlinear differential equations function similarly.) These equations are "recurrent" or "recursive," with "feedback loops." Because they reveal so clearly what happens when iterating such an algorithm, it bears pausing a moment to examine one of these impish formulas at play.

The case of population growth demonstrates aptly the play of a difference equation when it is subjected to the iterating power of a computer. Population biologists (and common sense, at least since Malthus) tell us that a population will increase at a rate of biological reproduction over a period of time and that it will be constrained by its environment during the same period. The rate of increase properly combined with the rate of constraint will describe accurately, therefore, the actual growth of a population. The trick lies with capturing these two rates in an equation that can yield the calculations.

We start by letting x stand for a population, say, of bugs in a pond, at a certain point in time. The bugs reproduce at rate R, so that R times x produces an unrestricted bug population of x_n, the next population, after an interval of time has passed. But bug reproduction also faces constraints (predators, lack of food, boating accidents). These we can note with the function $1 - x$, with 1 standing for the maximum possible bug population of the pond. The actual population is expressed as a fraction ranging between 1, the maximum, and 0, extinction. For instance, if 5 million bugs is the pond's supportable maximum, and it contains a million bugs, then $1/5$, or 0.2, represents the actual population. The $1 - x$ function expresses restriction to bug growth because as the fraction x increases, $1 - x$ becomes smaller. Combining now our rate of expansion, Rx, with the rate of constraint $(1 - x)$ we obtain the mathematical function, or $f(x)$, that simulates the population growth of bugs over time: $x_n = Rx(1 - x)$.

With this, rather simple, algebraic equation in hand, we can begin testing or experimenting with growth possibilities by substituting different numerical values for the rate of reproduction (R) and for the starting population (x). (Changes in R values are generally the more significant.) After each calculation, x_n simply becomes the new x and the calculation repeats. Remember, each iteration of the equation corresponds to one time interval in bug population growth or decline—day, week, month, year, and so on. One can figure these multiplications easily by hand or with a pocket calculator, but making 10,000

TABLE 9.1
Chaos and Complexity in Population Growth

Population growth is captured in the equation $x_n = Rx(1 - x)$,
where R = the biological reproduction rate of any population, such as bugs in a pond, and
where x = the starting population (expressed as a fraction between 0, extinction, and 1, the maximum population possible) at an initial point in time, and
where $(1 - x)$ = the constraints on population growth from predators, the environment, and the like, and
where x_n = the next population after an interval of time has passed.

Procedure: After each calculation of x_n, the result becomes the new x in a successive iteration. Repeat iterations until achieving a steady state, oscillating state, cycling through regular sequence, complexity, or chaos.

Table of select values for R and x (in most cases R is the more significant variable; hence the initial value of x can remain the same):

Value of R	Initial Value of x	Results
$R = 1.5$	$x = 0.5$	Steady state of 0.3333, the stable size of the bug population
$R = 1.9$	$x = 0.5$	Steady state of 0.4736
$R = 2.0$	$x = 0.5$	Steady state of 0.5
$R = 2.6$	$x = 0.5$	Steady state of 0.6153
$R = 3.1$	$x = 0.5$	The steady state has bifurcated into two stable populations; iterations oscillate between two fixed values, 0.5588 and 0.7645
$R = 3.4$	$x = 0.5$	Oscillating between two fixed values, 0.8421 and 0.4518
$R = 3.5$	$x = 0.5$	The two populations have doubled again; now iterations cycle repeatedly through a sequence of four fixed values, 0.5008, 0.8749, 0.3828, and 0.8269
$R = 3.55$	$x = 0.5$	Repeated cycling through eight different values (or stable populations)
$R = 3.57$	$x = 0.5$	Repeated cycling through 128 values

For R values from 3.6 to 4.0, results vary widely. Some iterations cycle repeatedly through extremely large collections of values, which are formed through additional doublings (128, 256, 512, 1,024, 2,048 . . .). Other iterations are chaotic, with no apparent repetitions through cycles at all. Still others remain on the border between regular cycles and chaos. The latter R and x values produce complexity.

$R = 4.0$	$x = 0.08$	Chaos: random values
	$x = 0.36$	Chaos: random values
	$x = 0.73$	Chaos: random values
	$x = 0.75$	Surprise! steady state of 0.75
	$x = 0.78$	Chaos: random values
$R = 4.1$	$x = 0.5$	Initially chaotic oscillations; leading to steady state beyond parameters

iterations of the same equation will take time. A computer does it in literally fractions of a second.

Table 9.1 shows an array of possibilities, ranging from simple order to chaos. After a few iterations, for values of R below 3, the populations settle down to a steady state, in which the number of bugs being born exactly matches those dying during the same period. Then, around $R = 3$, the first significant change in the pattern occurs. The steady state bifurcates, and the iterations begin to oscillate between two precise and stable populations. This pattern continues with R values in the 3.1 to 3.4 range. When we move to $R = 3.5$, the number of states doubles or bifurcates once again. Now, rather than oscillating between two numbers, the iterations produce a set of four different values, whose sequential pattern is repeated in cycle after cycle. (This set of values is termed an "attractor.")

After $R = 3.5$, the bifurcations occur ever more quickly with the slightest increases in R. Such period bifurcations or doublings were discovered by mathematician Mitchell Feigenbaum, one of the founders of chaos theory, who demonstrated both the universality of bifurcation patterns and their inevitable path to chaos in this and comparable equations. Figure 9.1 shows the doublings of our population equation on a typical Feigenbaum graph. As we increase R to $R = 3.55$, we see the next doubling, a cycle of eight values (2^3). Then, nudge R to 3.57, and the cycling proceeds through a whopping set of 128 (2^7) numbers, too many to detect visually on the scale of our graph. This means the bug population will pass through 128 different sizes in a cycle, then repeat the pattern for another 128, and so on. To this point, the patterns of the iterations have become more complicated, but they stay nonetheless stable, repetitive, and predictable.

Between $R = 3.6$ and $R = 4.0$, strange things happen to our numbers. Some values of R and x generate domains of the bug population that are still stable and ordered, but with iterations passing through huge, repetitive cycles (all of which are created by further doublings, powers of 2—128, 256, 512, 1,024, 2,048, 4,096, and so on). For other values the doublings themselves become chaotic, with no repeatable cycles and no predictable patterns in the bug population whatsoever. Yet other R and x values sit on the border between stability and chaos. Although not predictable, they do remain stable and viable over a long period of time. This behavior—stable, yet unpredictable—is called complexity. Mathematically speaking, complex number domains lie in the regions around bifurcation. Technically, therefore, the term 'complexity' signifies a "quantitative measure" that can be assigned to a physical or biophysical system or computation lying between "simple order and complete chaos," a territory "orderly enough to ensure stability, yet full of flexibility and surprise."

When we increase R to 4.0, the numbers become completely chaotic, fluctuating wildly with no repetitive patterns of cycling or predictable values. This is "deterministic chaos," so called because it is generated by following the logical rules of an algorithm as it proceeds through numerous iterations. Even here, as with most iterations of R values between 3.6 and 4.0, the results remain within our parameters of 1 and 0, the maximum population and extinction. This indi-

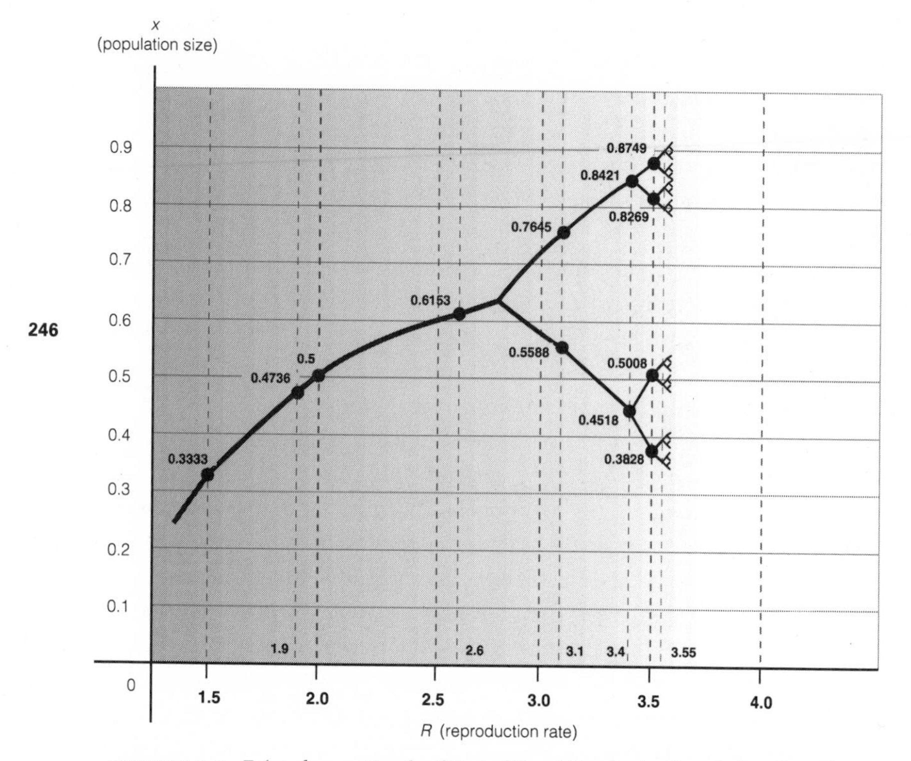

FIGURE 9.1. Feigenbaum Graph of Period Doubling in the Population Equation (Note: After $R = 3.5$, the doublings cascade rapidly with the slightest increases in R. From $R = 3.6$ to $R = 4.0$ results vary markedly, ranging from stable populations to chaotic, unpredictable ones. Within this range, domains of complexity are found at borders between cyclical stability and chaos; these are areas around the points of bifurcation.)

cates that the bug population, though disorderly or unruly or unpredictable, is staying quite viable. Then, finally, when numerical values for R exceed 4.0, the results show a quick trip to the boneyard: a steady state at either below extinction or explosively above the maximum.

Steady state, predictable oscillations and cycling, complexity, chaos—such are the possibilities emerging from computer play with this single, difference equation. Like the case of population growth, in instance after instance the results of devising iterating computer simulations of natural phenomena have been remarkable and have helped open up new vistas of scientific thinking. The play that constitutes information, in short, produces chaos and complexity not merely as isolated discoveries, but rather as new disciplines of scientific inquiry.

From these disciplines, their proponents argue, are surfacing new "deep structures" that can account for previously unexplained phenomena. "Structures" here are understood to be computational rules and the patterns they generate. In these disciplines too, creating computer simulations actually constitutes much of the research under way. (For this reason many experimental scientists remain skeptical of the more far-reaching claims made by the yea-sayers of the new fields.)

Progenitors of our age's computer power and play, both Alan Turing and John von Neumann had envisioned and contributed to the developments leading to complexity. In Turing's last scientific paper, he devoted his attention to a process known in biology as morphogenesis, the means whereby structure and function arise in living organisms. His approach was largely mathematical (or better, computational), and in effect he was proposing an early version of computer modeling to describe pattern formation in nature. Likewise, much of von Neumann's later work concentrated on "cellular automata," whose mathematical theory he created. A cellular automaton comprises an array of cells (such as a grid or checkerboard, with each square a separate cell). Each cell can be in a number of states (ON or OFF are the least complicated), and the entire grid can change according to preestablished, very simple rules. One can create different rules that govern the behavior of the cells and then watch what happens as the cells "evolve" over time through the program's successive iterations.

An extension of the Turing machine, cellular automata began in the 1950s as a mathematical curiosity that took advantage of a computer's iterating capabilities. A decade later mathematician John Conway created the "Game of Life," which took the play further and became the forerunner of a whole field of inquiry into "artificial life," a field devoted to computer simulations devised from cellular automata. Building on von Neumann's early ideas, Conway's computer "Game" employed a checkerboard grid, which filled the screen, and three simple rules, which controlled the ON/OFF status of all the "central cells" (those surrounded by eight others):

1. The cell will be ON in the next generation if exactly three of its neighboring cells are currently ON.
2. The cell will retain its current state if exactly two of its neighbors are ON.
3. The cell will be OFF otherwise.

Conway designed these rules to balance each cell's "life," to prevent its dying of isolation with too few neighbors or dying of suffocation with too many.

With rules in place, one can stipulate a number of ON cells at the start of the program, then let the computer take the grid through as many iterations, or time cycles as desired. At each cycle the entire screen changes in accordance with the three rules. As with the above population equation, in this exercise there emerge a plethora of various patterns, ranging from steady states to oscillations to complex and unpredictable arrays. Some patterns, tagged as "gliders," slide at a 45-degree angle across the grid; some, labeled "eaters," annihilate the gliders when they collide; other cells collide and create new patterns. To date, programmers have hacked out a large inventory of life forms, but the limits of the "Game" have yet to be reached. New patterns are still possible.

Behind these playful, cellular automata, as well as implicit in Turing's last work, lies a novel approach to many scientific issues and questions. The approach assumes (1) that the laws of nature function as algorithms and (2) that natural phenomena process information in a manner like that of a computer. (The same ideas underlie studies in AI, artificial intelligence, which consider the human brain to be an information processor, a computer.) Key to the sciences of complexity, these assumptions have opened the door to computer simulations as a new arena of research. The simulations themselves use algorithms that express directly the primitive 'and', 'or', and 'not' gates of Boolean logic, the computer's central processing components. Equally momentous, the automata driven by these playful algorithms sometimes generate totally new phenomena, properties, and structures—emergent systems. They transform computer play into an instrument of scientific discovery.

Emergence

In our bug equation, complexity refers to those populations remaining viable and stable over a long period of time, even though from cycle to cycle their exact size is unpredictable. Here, and more generally, complexity involves applying some quantitative measure to a range of phenomena, or to their computer model. But simulating complex phenomena can also lead to the discovery of new, emergent properties and structures, patterns that entail more than just statistical measures. Gell-Mann calls these phenomena "complex, adaptive systems." They are systems emerging from dynamical processes that are open-ended, that are not in a state of equilibrium, and that follow the arrow of time.

A crucial difference between mere statistical complexity and the emergence of new properties or structures resides with the presence of information processing through feedback. (The precise boundary between complexity and emergence is still much disputed; more on this below.) Some phenomena are now understood to be processing information constantly as they react and adjust to their environment. They assimilate or "learn" new information from circumstances. They use this information to adapt themselves to changing situations. In so doing, they alter their properties and even structures, sometimes forging entirely new ones.

These dynamical systems differ from other complex, often turbulent systems that can also be simulated in part by means of nonlinear equations but that do not process information. Water running at low pressure from a kitchen faucet, for instance, flows in a steady stream; increase the pressure and volume, and the stream moves from a steady into a complex, turbulent state. Though mathematically complex, turbulent water flow is not "adaptive"; it produces no new, emergent structures. In this and similar dynamical systems, no "information transfer" or "learning" is ascribed to the phenomena themselves.

When applied to adaptive systems, many of the statistical features of complexity, with their often quirky and surprising results, can help account for emergent properties. The so-called butterfly effect supplies one example. A colloquial metaphor, it describes a dynamical system's extreme sensitivity to initial conditions. The term comes from weather dynamics and refers to the compounded and exaggerated effects of a butterfly flapping its wings in Hong Kong on weather patterns in, say, Chicago. Mathematically, extreme sensitivity means that with some formulas tiny changes in initial variables can produce widely divergent results. (Note, for example, that in the above difference equation, a bizarre and exceptional outcome follows when $R = 4$ and $x = 0.75$; the population begins and remains in a steady state at its initial value. But alter the initial population by only the tiniest amount, and the consequences are again chaotic.) In adaptive systems sometimes these very slight, initial changes in information feedback may function like the butterfly's wings, compounding into complexity as the systems proceed through time. Under optimal conditions, the "information" fed back leads to emergent properties and patterns.

"Phase transition" describes a related behavior, the sudden transformation of phenomena into new structures or patterns, into new order. (The rapid crystallization of ice in supercooled water exemplifies a phase transition, from a liquid to solid state.) In adaptive systems, information steadily fed back into the system can sometimes reach and cross a statistical threshold, resulting in a nearly spontaneous appearance of new, complex order or properties. Like the butterfly effect, phase transitions can be simulated on computers. Scientists can "tweak" them (a favored synonym of "toy with") by changing variables to see under what specific conditions sudden transitions into organized complexity might occur.

Combining statistical complexity with information feedback, then, scientists use computer simulations to investigate the conditions under which new, organized systems emerge. Complex systems tend to dwell, optimally some argue, at the "edge of chaos." This is a zone where phenomena can be both stable and changing, where a judicious blend of stasis and adaptation makes survival most likely. Steady states (by definition) do not change; chaos (by definition) is random and structureless, even though produced by simple algorithms. Dwelling at the edge of chaos, complex adaptive systems are optimally suited for novelty, for emergence. This makes them the target of what writer and mathematician John Casti terms the new sciences of "surprise."

We can examine emergence as it rises out of the cellular play of complexity more concretely by following an argument of Stuart Kauffman, a theoretical biologist associated with the Santa Fe Institute. In the last three decades he has become a chief innovator and proponent of applying computer simulations to the study of perhaps the greatest emergence of all: life. Part of his work is avowedly "heretical," and Kauffman considers himself a "renegade" from mainline biologists (some of whom have labeled his studies, not as a compliment, "computational biology"). With such cautions in mind, however, we see that he argues a rigorous and incisive case to the scientific and general public alike.

Kauffman wants to amend seriously the Darwinian theory of evolution with accounts of the "self-organization" of phenomena, what he terms "order for free." Standard biological thinking generally focuses on the nucleic acids, DNA and RNA, as the chief perpetuators of individuals and species down through the ages by means of the "magic of template replication," the engine of natural selection. But natural selection cannot explain how these genetic building blocks of inheritance came about in the first place. Rather, it assumes them. Nor can the standard view account for cases where small changes in a population (whether of molecules, cells, organisms, or even species) have resulted in sudden transformations, for the good or ill of the population. This is because the standard view also assumes "gradualism," the belief that all features of complex organisms evolve over time only by gradually culling out and accumulating useful variations from among a cornucopia of random mutations. Evolution, Kauffman believes, is much more than mere "chance caught on the wing," as it was lyrically described by Nobel biologist Jacques Monod.

At the heart of his argument are "Kauffman networks" (so named by others), complicated cellular automata, designed to mimic complex chemical and biological processes. With them Kauffman seeks to explain how life on earth might have emerged and evolved from the "chemical soup" that made up our planet in

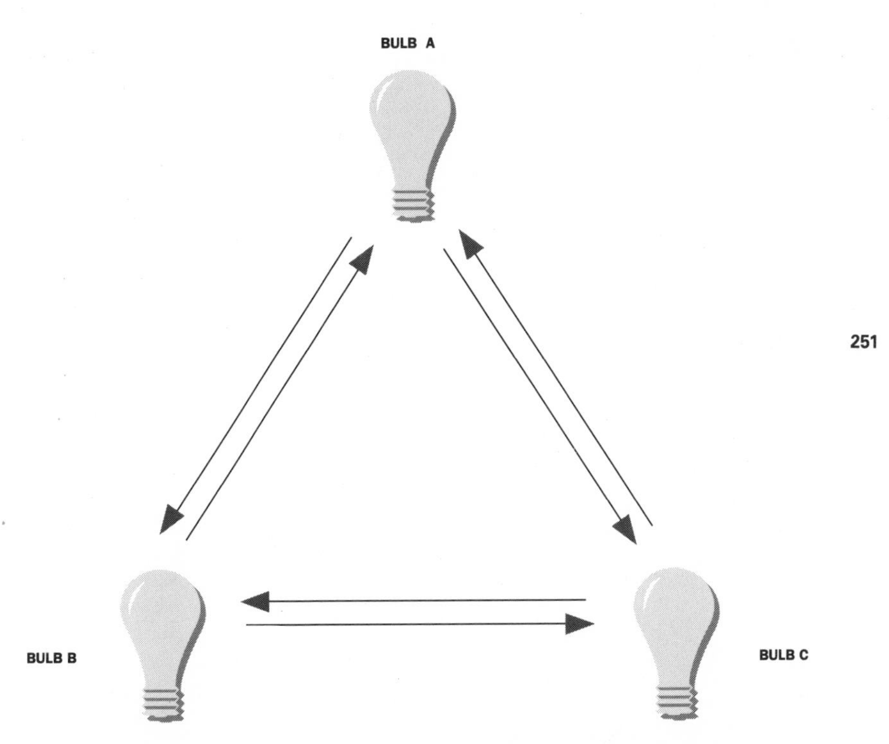

FIGURE 9.2. Lightbulbs Wired in a Boolean Network as Three Binary Elements

its early billennia. These automata are based on Boolean sets, much like those we have already encountered. They generate "computer-simulation movies," which among other applications yield a simplified idealization of how, in primordial times, molecules might have bonded chemically to create life through a process known as autocatalysis.

By itself, catalysis means the acceleration of a chemical reaction or bonding between two (or more) molecules, which takes place in the presence of a third molecule, a catalyst such as an enzyme. Say that under normal conditions, molecules A and B combine over time to make a third molecule, C. Now add the presence of a catalytic agent, molecule D, and the entire bonding reaction between A and B to create C speeds up markedly. Autocatalysis occurs when the creation of C serves, through feedback, to catalyze further bondings between A and B. Kauffman hypothesizes that in the right circumstances autocatalytic bondings can generate organic, "spontaneous order" among molecules and eventually cells. (The aggregation of organic slime molds on a pond's surface supplies a familiar instance of autocatalysis.)

Bulb A, AND:

B	C	A
0	0	0
0	1	0
1	0	0
1	1	1

(1 = ON; 0 = OFF. When B and C are both ON, then A is ON; in all other cases A is OFF.)

Bulb B, OR:

A	C	B
0	0	0
0	1	1
1	0	1
1	1	1

(When either A or C is ON, then B is ON; when A and C are both OFF, B is OFF.)

Bulb C, OR:

A	B	C
0	0	0
0	1	1
1	0	1
1	1	1

(When either A or B is ON, then C is ON; when A and B are both OFF, C is OFF.)

FIGURE 9.3. Rules Governing the Boolean Network Depicted in Figure 9.2

To see how this self-organization comes about, begin with a small Boolean set of three binary cells or elements, say, light bulbs, that could be either ON or OFF. In this set, the bulbs are wired so that each can provide an input signal to the other two and likewise can receive an input from each of them, as indicated in figure 9.2. The input signals to each bulb may be ON (1) or OFF (0), and with this arrangement one can specify a bulb's response, making it depend on the signals it receives. Four possible pairings of input signals exist for each bulb: 00, 01, 10, and 11. If we let bulb A light only when it receives an input (1) from both B and C, then it represents the logical 'and' gate. Similarly, if we let bulbs B and C light up with a signal from either or both of the others (01, 10, 11), they stand for the logical 'or.' As figure 9.3 shows, these possibilities match the (by now) familiar Boolean lattices for logical multiplication and addition that we diagrammed in chapter 7.

Notice in this example that there are three bulbs and two inputs to each bulb, making eight possible states, or combinations of the three bulbs ($2^3 = 8$). Figure 9.4 displays the entire array of possible states at a given time *T*. Now, we activate the system by turning on the bulbs, and then note which other ones

	T			T+1		
	A	B	C	A	B	C
Row 1	0	0	0	0	0	0
2	0	0	1	0	1	0
3	0	1	0	0	0	1
4	0	1	1	1	1	1
5	1	0	0	0	1	1
6	1	0	1	0	1	1
7	1	1	0	0	1	1
8	1	1	1	1	1	1

Column T represents all the possible ON/OFF states of the three bulbs, A, B, and C. Column $T+1$ indicates the ON or OFF state of each bulb after it has been activated at T in accordance with the Boolean gates listed above in figure 9.3. Read across in rows. In row 7 at T, for example, bulbs A and B are ON, while C is OFF. At $T+1$, the subsequent moment, A is OFF, with B and C ON. In the next iteration we substitute the results at $T+1$ for a new T, and then read across in row 4 to obtain the states of the bulbs at $T+2$. From Stuart Kauffman, *At Home in the Universe*, p. 76.

FIGURE 9.4. Iterating Possible States in the Boolean Network

light up in the next moment of time, at $T+1$. For each state (rows 1 through 8), the logical combinations of ON/OFF inputs from the bulbs at T determine the ON/OFF states of each bulb at $T+1$. These combinations are governed by the (arbitrarily) assigned Boolean functions listed in figure 9.3. The table in figure 9.4 gives us the results. Take row 7, for example. The B (ON) and C (OFF) bulbs in T determine that the A bulb in $T+1$ will be OFF. So too, still in row 7, A and C determine B, while A and B determine C. Now we substitute the results at $T+1$ (A-OFF, B-ON, C-ON) for a new T. Reading across in row 4 (0, 1, 1) we see all three bulbs ON at the subsequent moment, $T+2$. After that (see row 8), we shall have a steady state, with all bulbs staying lit.

This tiny cellular automaton illustrates the general procedure Kauffman follows in developing much larger and more interesting networks of elements. By substituting different values for the number of bulbs (N) and for the number of inputs to each bulb (K), and then letting the computer iterate the computations over time, strange patterns and results appear in the form of blinking lights, along with the statistical data on hard-copy printouts. As with other examples of automata and with the population equation, these patterns range from steady states, to oscillations, to complexity, to chaos.

Kauffman "plays" (his term) with these simulations to find patterns of self-organization that would model the autocatalysis leading to life. (In fact, he engages in further mathematical maneuvering, which reaches beyond present concerns, to complete the models.) Generally, at lower *N* and *K* values, like the above case with three bulbs and two inputs, the networks become quickly frozen into a steady state, and the same bulbs alternate or remain on, cycle after cycle. No life here. In some larger sets, where the number of bulbs equals the number of inputs to each bulb ($N = K$), the results become thoroughly chaotic, with bulbs blinking in wild abandon. Here is too much life, with no homeostasis, the order and stability needed to ensure the survival of the molecular population.

Order—"sudden and stunning"—arises out of the experiments in $K = 2$ networks with large populations. Although they initially look like a mad hatter's jumble of wires and bulbs and frenetic randomness, these networks prove to be "well behaved" and settle into complex, but life-sustaining homeostasis. Let $N = 100{,}000$ and $K = 2$. Imagine 100,000 light bulbs wired randomly so that each bulb receives inputs from only two other bulbs. Imagine, further, that each bulb has been assigned at random a Boolean function, such as 'and' or 'or'. The system has $2^{100{,}000}$ possible states. Even at a nanosecond per iteration, cycling through this number of states would take vastly more time than exists in the entire life of the universe. But what happens? The network soon calms down and begins to cycle among the square root of 100,000 states, a "mere" 317 possibilities. Thus complex, though unpredictable, order emerges from 100,000 randomly connected bulbs. After years of experimenting with different values for inputs and networks, Kauffman has found that sparsely connected networks (those where $K = 1$ or 2) exhibit a great deal of internal order and stability, whereas densely connected ones, where *K* approaches 4 (remember our population equation), border on and produce chaos.

What do these blinking lights mean for the question of whence life on earth? All matter is made up of atoms bonded together to form molecules, such as the familiar H_2O, a pair of hydrogen atoms united with one of oxygen to constitute a molecule of water. Through further bondings, molecules create chemical compounds, which convention divides between organic (involving carbon) and inorganic (excluding carbon). Somewhere around 3.5 to 4 billion years ago the molten mass of matter that became the earth cooled sufficiently to support liquid water, as opposed to vapor. This was the original "chemical soup," and it contained all the basic elements: carbon, oxygen, nitrogen, hydrogen, and others. Water provided the arena where these elements jostled one another. Using energy from the sun and electrical storms, they united rather quickly (within some 300 million years after cooling) into different organic compounds. (Scientists now count over a million of these compounds.) Among the earliest were amino acids, the backbone of proteins. Proteins, in turn, tendered the scaffolding for cells, as well as the enzymes hastening further chemical reactions. Some, nucleoproteins, bonded with other organic compounds, the nucleic acids RNA and DNA, to become the chemical basis of heredity and natural selection. Voilà, life.

The foregoing sketches a plausible scientific speculation, not a confirmed scientific account, although portions of it are quite firmly established. So too with Kauffman's computer accomplishment. Roughly, the inputs in the models represent chemical bonding processes (which are, to be sure, much more complicated than a simple ON or OFF), while the bulbs stand for populations of different molecules. The complex order that emerges suddenly in the $K = 2$ variety of simulations plausibly matches early, "self-reproducing chemical networks," a "proto-cellular" world whose metabolisms are "whole and complex," and whose "ecosystem" strikes a viable compromise between stability and malleability.

Kauffman takes pains to keep his "movies" within the parameters of known chemical and biological processes, fossil records, and other evolutionary evidence. And he insists we simply do not yet know which and how many molecules bonded autocatalytically to seed the evolutionary tree of life. These precautions notwithstanding, the lessons of his computer play are profound ones. Simulations demonstrate that not just any old chemical soup will do. Too few molecules and bondings go nowhere; they just lie there, inert. Too many molecules and bondings yield chaos, with no life-sustaining order. Too many molecules without autocatalysis would never have enough time to make life together; the universe would die first. Yet, gradually increase molecular diversity and complexity over the span of a few million years, and autocatalytic feedback begins. Upon reaching a threshold (a phase transition), complex, self-reproducing networks suddenly appear. Critically too, the time frame for these events matches the best evidence and estimates of life's advent. The earliest known fossils, representing what experts believe to be well-formed cells, date from 3.437 billion years ago, roughly 300 million years after the arrival of liquid water. In this period molecules would have had ample time for the random opportunities required to find autocatalytic bonding partners and create self-organized and sustaining metabolisms.

Extending his simulations of self-organization to other, key biological processes and evolutionary moments, Kauffman models the emergence of such phenomena as the following: "a genetic regulatory system"; our biosphere's "stunning molecular diversity"; the "burst of evolutionary creativity" after the

Cambrian explosion of 550 million years ago and again after the Permian extinction, 245 million years ago, when 96 percent of all species disappeared; cell differentiation and morphogenesis (the "mystery of ontogeny"); and even "global civilization," at whose core resides "pluralistic democratic society . . . part of the natural order of things" (on which more in our conclusion). At this juncture, we must forego any further foray into these intriguing topics, only reaffirming in summary that from the play of computer simulations has sprung a new orientation to some of the most challenging questions concerning our ever changing, evolving, and emerging world.

As we have seen, information play in complexity and emergence relies on algorithms. These are logical, sequential instructions given to the computer, providing it the digital strings of information to process or compute, and the order in which to proceed. The steps of an algorithm are carried out one after the other in successive moments of time, however tiny such moments might be. This holds even when repeating computations, for each new iteration processes the information produced by the previous cycle. What scientists call the "arrow of time," the irreversibility of time's flow, is thus built into information play. When the laws of nature are understood as algorithmic, capable of being simulated by a computer, they too assume and exhibit the arrow of time.

Stated another way, in computer simulations time is asymmetrical; it has direction. This makes time constitutive of information in a way it was not for the classical or modern periods. With analysis, for instance, modern science addresses the question of how things function. Mathematical "laws" describe in formulas the functional dependency between components of phenomena—units of force, mass, movement, position, and the like. Here, time is "symmetrical." It acquires direction only when initial and boundary conditions are established and the formula's functions are set in motion. No directional time flow enters the functional relations themselves. Planets could orbit their sun in a clockwise way or, "backwards," going counterclockwise, and the law of gravity would remain the same. Thus does modern, reductive science (including relativity and much of quantum mechanics) treat time in a manner that rubs against the grain of our ordinary sense of its irretrievable passing.

With complexity and emergence, time's asymmetry unveils a recasting of scientific questions. Instead of asking how phenomena function, simulations reveal how phenomena come about. Of course, the latter question has long been a central issue in biology ever since Darwin (and even before), and in other sciences as well, especially thermodynamics. But the abstract and computational rigor now made possible by computer technology has invested these questions with a new richness and complexity of their own. Algorithms only run "forward," relentlessly and unpredictably forward.

The Play of Pure Technique

The sciences of complexity are still in their infancy, which helps explain why much of their terminology has not attained the definitional rigor of older, more established portions of our scientific knowledge. Vague terms reveal that a clear boundary between complex and emergent phenomena has yet to be determined. Moreover, tentative expressions often riddle the accounts in these sciences. Kauffman, for instance, describes phenomena frequently with "if . . . then," "appears to be," "suppose that," "might be," "perhaps is," and other, subjunctive locutions. Nor does he stand alone in this linguistic looseness, which further marks the play of contemporary information.

In part, linguistic looseness derives from the novelty of applying computer simulations to phenomena that crisscross and overlap traditional disciplinary divisions (such as between the physical and life sciences). In part, too, it stems from stretching the metaphorical reach of terms such as 'complexity', 'learning', and even 'information' itself. Not surprisingly, at times these figurative expressions can become quite muddled or even circular. What sense do we attach to the claim that a paramecium "learns" from its environment? Learning, we read, means taking in "information" and using it. But then by what token do we identify information, if not as the content taken in or learned?

The tendency to invoke 'information', 'learning', and related terms loosely and even circularly in describing all sorts of phenomena not only reveals the idiom's ambiguities. It also suggests that, as metaphor, information rhetoric cannot keep pace with the power and play of today's information technology. In fact, the lag between the fast-paced technology and our grasp of it in image and word contributes much to the general, confused hype about information. (Recall similar discrepancies between the abstractions of modern, mathematical analysis as an information idiom and the often oxymoronic, conceptual efforts to describe its contents.) Today's idiom uses algorithmic rules to surround or bracket phenomena by means of simulations, a bracketing that constitutes information's pure technique of power and play. Simulations stand in place of the rhetorical definitions or of the linear, predictive formulas of earlier ages. Rhetorically, the term 'information' has become simply a placeholder indicating specific algorithms at play in a simulation. The term's vagaries, therefore, derive from the wealth of possible simulations, whose horizons recede with the advancing technology.

Yet a reason deeper than mere linguistic looseness intrinsically yokes play to the new sciences. Beyond the confusions of rhetoric, as we have suggested, an inherent indeterminacy marks complexity. To study complex phenomena, Gell-Mann observes, one must always specify the level of detail under examination, a procedure termed "coarse graining" in physics. "Any definition of complexity," it follows, "is necessarily context-dependent, even subjective." The subjectivity extends all the way to sophisticated, computational definitions of "algorithmic complexity" (also named "algorithmic information content" and "algorithmic randomness"). These are the quantitative measures of digital strings that contain information messages. Because no algorithm can determine in advance the length of these strings, they always contain some built-in randomness (remember Turing's "noncomputable" number). The actual information content can be established only by letting the computer iterate or cycle—by letting it play—to produce the varying results shown by Kauffmanesque models. In the strictest and most rigorous sense, therefore, with complexity not only is information undefined, it is indefinable, indeterminate.

Propagated by the algorithms of computer simulations, the indeterminacy of information is woven into the very fabric of our contemporary age. As Kauffman notes, simulations explain; they do not predict. The results are "deterministic" because they are indeed "determined" by the algorithmic rules that logically govern the computations at each new iteration. But they are not "predetermined," for precise quantitative values cannot be predicted linearly from the equations or algorithms themselves. Of course, we find patterns in the simulations. We expect the onset of chaos as R approaches 4 in the population equation, and Kauffman knows that his $K = 2$ networks will settle down into organized complexity at the edge of chaos. In this sense we can forecast general results of simulations, but we cannot predict them precisely.

Indeterminacy in the sciences of complexity signals the absence of closure in today's information technology. Closure, we noted earlier, allows us to circumscribe our natural, human drive for order, which might otherwise degenerate into compulsive, Rabelaisian list making. It prevents us from being overwhelmed by the information we perceive. Both the classical and modern information idioms had provided closure through classifying and analyzing information. The contemporary idiom serves no such function.

We must take care here not to run amuck with the implications of this observation. Loss of closure neither symbolizes nor sanctions the "anything goes" of cultural relativism. Even less does it directly imply our being overwhelmed by experience, although we may well feel our age's information overload. Rather, it simply means the displacement and supplanting of well-delineated information structures with the open-ended, unlimited activity of information processing. Just as closure through analysis converted information to knowledge in the modern age, so too does computer simulation convert it to play.

Play is "freedom," wrote Huizinga, yet it also "creates order, *is* order." In its creative activity, play does not imitate or reflect or correspond or map directly to an outside world, although the resulting order it produces might eventually do so. Play's own order is not derivatively *mimetic,* but sui generis, *methectic,* as the Greeks would say in reference to the theater, a "helping-out of the action." By now it should be abundantly clear that information in the contemporary idiom means process, always in motion, always abetting the action of life's drama. Less clear, perhaps, this motion has direction, not toward any telos, purpose, or end, but following with its rules the arrow of time. From the time-bound movement of information play there can emerge novel, unforeseen structures, "order for free."

CONCLUSION

The Two Cultures and the Arrow of Time

> We shall not cease from exploration
> And the end of all our exploring
> Will be to arrive where we started
> And know the place for the first time.
>
> —T. S. ELIOT, "Little Gidding"

Writing in the twilight of the modern information age, novelist and physicist C. P. Snow lamented a disturbing development in the intellectual life of the mid twentieth century. Whereas intellectuals had once thought of themselves as a single group characterized by diverse interests, with scientists well versed in the humanities and humanists knowledgeable in science, Snow observed that they now tended to align themselves into distinct camps. Not only did humanists and scientists increasingly talk past each other but, he deplored, they also increasingly disdained each others' intellectual efforts. Snow questioned the long-term vitality and viability of an intellectual world thus divided.

The dawning of our contemporary information age may yet show the way toward a reconciliation, by revealing the distinctive role of the arrow of time in the sciences of complexity. Traditionally, scientists have prided themselves on seeking the eternal and immutable in nature, in contrast to the contingent and temporal concerns of the humanists. Now, with the aid of computer technology, scientists find themselves exploring in nature the very dimensions of contingency and temporality that had once distinguished them from the humanists. And as they apply the arrow of time to the study of nature, they discover new dimensions in phenomena, opening unexpected vistas of inquiry that rival those of the reductive sciences. The future may promise greater intellectual harmony as both camps, while still remaining separate, follow increasingly parallel paths, seeking in their respective ways to understand time's irreversible processes.

Yet, as we enter the contemporary information age, we bring the implacable hostilities of late modernity with us, threatening the potential for reconciliation. Each of Snow's two cultures has seized upon the contemporary importance of the arrow of time as a weapon to employ against the other. In the humanistic camp, some "critical theorists" in the academy have lately begun to use the arrow of time to relativize traditional science, to reduce its laws to social constructions. These critics disparage scientific claims to objectivity as examples of rigid, outmoded, "linear" thinking, as opposed to the kind of imaginative, avant-garde, "nonlinear" thinking that reveals the conditional, provisional, "emergent" quality of all truth claims.

Though imprecise and slippery, these metaphors from the sciences of complexity are among the more accessible ones in the linguistic arsenal of critical theory, whose literature is notoriously dense, forbidding, and sometimes even perversely incomprehensible. (One physicist, confronting the prospect of death, is said to have consoled himself with the thought that at least he would never again have to look up the word 'hermeneutics'.) Caricatures of their impenetrability notwithstanding, the structuralist, poststructuralist, and deconstructionist critics aim with utmost seriousness to free us from an insidious enslavement to socially constructed systems of meaning. Chief among these is modern, reductive science, which (they say) creates the reality it purports to describe. By attacking this very citadel of Western rationalism, the critical theorists want to goad us into viewing the world afresh.

Despite its salutary aspects, this critique tends by its very nature to go too far, violating strictures of common sense, culturally conditioned though these may be. Philosopher W. V. Quine puts this relativizing view of science in its proper perspective. References to things ostensibly outside language must remain "inscrutable," he argues, and things themselves can only be grasped as relative to the language, natural or symbolic; in which they are cast. In his view the "gods of Homer" and the "physical objects" of contemporary science share the same "epistemological footing" as "cultural posits," differing "only in degree and not in kind." But, he continues, even though gods and physical bodies (or matter) may stand equally as cultural posits, equally mythic, the "myth of physical objects" nonetheless generally proves "more efficacious . . . as a device for working a manageable structure into the flux of experience." We would do well to believe in this cultural posit, rather than in the intervening powers of Zeus and his cohorts, for it enables us to structure our experience consistently, reliably, and predictably.

Whereas critical theorists use the new information idiom to subsume traditional science under their own, relativistic views, those at the forefront of the sciences of complexity are doing the reverse, extending the idiom into the traditional preserves of the social sciences and humanities. In a field like economics, which seeks rules governing a circumscribed, quantifiable range of human behavior, this move makes perfect sense, enabling researchers to simulate the flow of money in a national economy or the fluctuations in a stock market. But such applications have also encouraged more grandiose visions of a reduction of the human world to the sciences of complexity, thus posing a second, imperialistic threat to the reconciliation of the two cultures.

The formidable Murray Gell-Mann, for example, touts the sciences of complexity as a means of understanding not only biological evolution, where experience is compressed in DNA, but also social evolution, where institutions, customs, traditions, and myths comprise kinds of "cultural DNA," interacting with the environment in the manner of complex, adaptive systems. In a similar vein, biologist Stuart Kauffman claims that the sciences of complexity offer "fresh support for the idea of a pluralistic democratic society, providing evidence that it is not merely a human creation but part of the natural order of things," an emergent structure at the "edge of chaos." In contrast to the crabbed relativizing of the critical theorists, the threat to cultural reconciliation posed by the proponents of computer modeling is all the greater for being born of a brash, infectious enthusiasm for powerful new techniques.

When extended to the realm of politics and culture, the contemporary information idiom becomes a metaphor that masks a conceptual segue between incommensurable phenomena, from the complexity of natural systems to the historicity of human ones. Like emergent natural phenomena, historical phenomena are exquisitely sensitive to initial conditions, following the arrow of time into an unpredictable future. What better evocation of the butterfly effect than George Herbert's famous line "For want of a nail the shoe is lost, for want of a shoe the horse is lost, for want of a horse the rider is lost"? And if that rider were a king, say, Shakespeare's Richard III ("A horse! a horse! my kingdom for a horse!"), we could well imagine the importance of a nail in the grand scheme of things. Momentous human events often hinge on factors almost as trifling as the fluttering of a butterfly's wings. But once the current of historical causation is set in motion, does it follow algorithmic rules of emergence? Do its coursings fall within the statistical range of complexity? The information metaphor obviously breaks down. Whereas the emergence of complex, adaptive systems is characterized by a deterministic, rule-driven unpredictability that admits of statistical description, the course of historical development is simply unpredictable.

Smart people are seduced by this information metaphor because it promises to fulfill the social scientific and, ultimately, Cartesian dream of discovering the algorithmic-like rules governing not only our physical but also our social existence. Thus Kauffman concludes *At Home in the Universe,* his fascinating exploration of self-organization and complexity in biological systems, with a coda on the "emerging global civilization," in which he speculates that the sciences of complexity point the way toward a "post-Marxist analysis of social evolution," revealing the proper role of "law" in the "historical sciences":

> Is there, in fact, a place for "law" in the historical sciences? Is the Industrial Revolution, with its explosion of technologies, an example of assembling a critical diversity of goods and services, of new production technologies, such that the diversity fed on itself autocatalytically? What of the cultural revolutions such as the Renaissance and the Enlightenment? Do these somehow reflect a web of collectively self-reaffirming ideas, norms, and agencies?

In place of the nineteenth-century historicism of Hegel and Marx, Kauffman envisions a history informed by the sciences of complexity, a history of self-organizing structures emerging at the edge of chaos.

But this kind of complexity-based history is only historicism recycled through a new vocabulary, one replete with metaphors drawn from the new information idiom. Thus the "emergence of complex, adaptive systems" stands simply for the development of historical entities, "phase transitions" for seminal events and movements, and "feedback loops" for the effects of historical awareness. Good heavens, to paraphrase Molière's would-be gentleman, our computational scientists have been "speaking history" all this time without even knowing it!

Of course, they might still object that they seek the computational rules and patterns underlying historical phenomena. But let them attempt a computer simulation of the Renaissance, Enlightenment, or Industrial Revolution—as though such protean, historical "periods" were objective entities, with readily identifiable and quantifiable characteristics—and such claims will quickly go by the board. Then even our computational scientists will acknowledge that the only proper simulation of historical events and entities is the story of how they came about.

Working from within the contemporary information idiom, Gell-Mann, Kauffman, and others have extended asymmetrical time and the sciences of complexity to embrace historical as well as natural processes. Yet in the segue from algorithm to history they have inverted the flow of argument, cart before ass, and misconstrued the context in which their work has arisen. For history is not a function of algorithms. To the contrary, algorithms—and the contemporary information technology whose endless processing they embody—arise from the very historicity of information itself. This is the story we have just told.

Our history of information has been the history of a type of abstract thinking, but not (how can we say it otherwise?) taken in the abstract, for information is historically grown. At each crucial point in its narrative, this type of thinking has been shaped and channeled by technological factors: literacy in all its forms, modern numeracy, and that logical by-product of numeracy we call the digital computer. And in their applications and consequences, these technological factors are themselves culturally and historically conditioned. The nexus of abstraction and technology that constitutes information is, to repeat, profoundly historical.

The fundamental fact of information's historicity liberates us from the conceit that ours is the information age, a conceit that underlies Kauffmanesque inferences from "computer-simulation movies" to history. It allows us to stand outside our contemporary information idiom, to see where it comes from, what it does, and how it shapes our thought. Likewise, the historical viewpoint enables us to step inside other idioms, to see how they functioned in their respective ages. It enables us to move between and among idioms in ways the idioms themselves do not permit. In so doing we see ourselves and our world from a critical perspective, from a vantage point that reveals an unprecedented opportunity for healing with historical thinking the historically rooted rift between the sciences and the humanities, a rift that extends back to Descartes's failed attempt to restructure human knowledge on the basis of numeracy.

As a means of capturing and expressing the historicity of our subject, we have referred throughout our essay to the ages and aging of information. The historical ages of information—classical, modern, contemporary—refer both to past eras and to their coexistence in the present day. Once in place historically, the information technologies of all three ages have continued to serve as modes of organizing human experience, of imposing order on flux, of shaping our interactions with nature and with one another. Each of these information idioms arranges our experience somewhat differently from the others or, perhaps better stated, shapes a different portion of it. Accordingly, once having arisen in historical context and sequence, the three ages of information can be said to coexist and to complement one another. We live in the contemporary age, yet the classical and modern ones surround us still.

The wisdom captured and extolled by means of the alphabet and the written word continues to resound with us, as does the ordering of practical affairs through classification. From the Greeks forward, alphabetic literacy has enabled us to use nouns and adjectives to identify and organize the various objects of our experience. Classificatory systems pigeonhole words and things, and in so doing answer the question "What?" What does it mean to be a particular kind of thing or object? What are its general and specific characteristics?

Likewise, the knowledge born of modern numeracy—the mathematical, reductive sciences—continues to inspire and to inform, as large segments of the now cosmic map fall within the scope of the analytical vision. This information idiom answers the question "How?" How do things function or work? How do variables and constants depend on one another? And this kind of question reveals how things can work for us. We fly to the moon not at dusk on the soft wings of Minerva's owl, but in the broad daylight of hard, mathematical physics.

As contemporary information technology extends its digital power and play over our lives, it facilitates answering these two kinds of question. Computers enable us to classify willy-nilly whatever words, things, and phenomena we so choose, either processing rapidly the conventional information of practical matters or inventing new information forms, new ways of classifying. Further, as high-speed adding machines, computers also aid in the ongoing, analytical project of reductive science. The real novelty of digital power and play, however, resides with the time-asymmetrical, iterating capabilities of computers, and their applications in simulating phenomena. Here complexity and emergence testify eloquently to the formulation of a different kind of scientific question, "How do things come about?" How do existing phenomena lead to emergent, new properties and structures?

What are phenomena? How do they function? How do they come about? These three questions reveal in its fullness the human compulsion for ordering information, for organizing experience. At times the questions stand in conflict, with the information from one idiom seemingly at odds or incompatible with that of another. (The classical age thus defined gravity as the quality belonging to the class of heavy objects, those that fall toward the center of the earth. The modern age recast it as a quantitative, functional relation, the product of a body's mass times its acceleration.) Even so, there is clearly a place in our experience for all three modes of information organization. Just as science and religion, once bitter foes in the cultural wars of the modern period, have finally come at least to recognize that they ask different questions, so too is it apparent that different information idioms respond to different informational needs. And we need to embrace all three questions and all three ages in order to connect ourselves fully and authentically, we might even say graciously, to the natural and human worlds.

As a verb, 'ages' designates the historical processes that situate our three cultural moments in a series of trajectories—reflection, abstraction, displacement—which themselves manifest dimensions of our historical understanding. Reflection stands as the most obvious form of this historicity. It simply means the natural propensity of the mind to rework and reshape the products of its own creation, to see its own abstractions from a critical perspective as the objects of further study, analysis, and organization. As we have amply chronicled, this tendency helps explain developments within an information technology—literacy, numeracy, and computers alike. And it also helps explain the outgrowth of computer technology from the analytical vision.

Reflection leads, unbeckoned, to a deeper dimension of information's historicity, the pattern of its ever-growing abstraction. As information has aged, our ways of connecting to the world have become increasingly attenuated and distant. Words refer to things; formulas map relations; digital strings encode processes. From referring to mapping to encoding, our ways of informing our experience have become further and further removed from the immediacy of that experience.

At the dawn of information, writing provided the first means of abstracting, of drawing away from the "here-and-now" flux of experience and taking mental contents from it, a process that manifested itself in the linguistic propensity to classify. The Greek alphabet subsequently realized this classifying potential to its fullest extent, contributing to the birth of science and philosophy. Aristotle's logical system carried the impulse to abstract and classify all the way to the empyrean heights from where its wisdom commanded Western intellectual activity for two millennia.

But as Alfred North Whitehead once noted, for all its utility, classifying stands as only a halfway point on the path to the greater abstractions of mathematics. Historically, the rupture of an overtaxed, classifying mind-set coincided with the emergence of symbolic, relational mathematics to create a new vision of information articulated in the language of analysis. Gradually the new idiom began to displace the classificatory one, especially in the realms of astronomy and physics where it reshaped knowledge of the natural world. The encyclopedists extended the idiom globally, insisting that to know phenomena—natural or human—with any measure of assurance, one must reduce them to analytical abstractions, not subsume them under taxonomies.

The displacement of wisdom by knowledge culminated in the scientific determinism of the nineteenth century. Let the biologists muck about in their "tangled bank," gathering bugs, worms, and other creepy-crawlies, and putting them into their taxonomical cubbyholes; differential equations will give us the hard-core truth about the world. Thus spoke the determinists. At the same time, however, the attenuations of mathematics and physics joined the abstractions of the newly formed symbolic logic to uproot analysis, to sever its techniques from its foundations.

With the marriage of symbol and circuit in the twentieth century, our contemporary information age has reworked the techniques and products of analysis and has taken abstraction a giant step further. It has done so through a mind-bogglingly simple recasting of information as pure technique, as the manipulation of nothing more than 1s and 0s according to some elementary logical rules. At the same time power and play have also injected directional time into the heart of our exchanges with the outside world. Consequently, today's abstract, digital strings paradoxically remove us farther from the immediacy of our experience than we have ever been, even while the arrow of time brings us closer to it than at any moment since the dawning of information itself.

Following information's history thus has led us back to the present, where we began, to our contemporary age and idiom. But we return home with a difference. Our own age stands as the pinnacle of abstraction in the historical development of information. Yet, embedded in these abstractions, asymmetrical time and emergence in the sciences of complexity bring us closer to our basic, reflective nature as the species that understands its own experience in time-bound process. This is the process whose future cannot be predicted with precision and whose inevitable passing produces whole new structures, historical as well as natural. Its time parallels and resonates deeply with our own, internal "sense" of life's inevitable march forward as it unalterably wends its way toward an untimely end. As today's sciences grow more developmental in nature, we are brought ever closer to our existential center as contingent, historical beings.

The arrow of time thus reveals parallel trackings in both the sciences of complexity and in historical understanding, trackings that course through the historicity of information itself. These trackings may well allow humanists and scientists alike to recognize the complementary nature of their cultures, to acknowledge the legitimacy of their respective curiosities and interests, and to find increasingly common areas for discussion. It used to be claimed that the nineteenth century witnessed the flowering of historical consciousness; might we speculate that the twenty-first century will be the real era of historical understanding by virtue of reconciling the two cultures?

Finally, the historical process of abstraction also embraces what we have referred to as the displacement of one age by another. The displacement of wisdom by knowledge and of knowledge by power and play bespeaks a direction in the broad sweep of our intellectual and cultural life, not necessarily a progressive one to be sure, but one itself impelled, inexorably it would seem, by time's arrow. In ever soaring abstraction, displacement discloses a twofold movement of subsumption and exchange. Generally, each new, more abstract idiom subsumes much of the previous age's information, such as when logical calculus symbolizes nouns and adjectives, or when digital computers solve differential equations. At the same time, this process involves the exchange of one idiom for another, as each information age coalesces around its own distinctive set of questions, assimilates and reshapes materials from previous ages, and pushes aside as irrelevant what lies beyond its own idiom. Displacement thus calls our attention to the continuing evolution of new and different means of informing.

A term of historical comprehension, displacement brings, if not closure, at least a fitting end to our essay and to our reflections. It makes room for other readings of our collective past and frees us by suggesting our own displacement in a future as yet unknown. Indeed, were the reductionist project of the analytical vision somehow fulfilled, were the "dreams of a final theory" to come true, we now see that entire kingdoms of information—the realms of complexity and emergence—would remain outside the boundary of its global map, territory still to be explored. The arrow of time in contemporary information unveils to us the fundamental, historical reality of human understanding, at once our grace and our redemption.

NOTES

(We have cited below only the sources of the most significant direct quotations. A general discussion of the scholarly literature on which we have relied follows in the Bibliographical Essay.)

Chapter One: Orality and the Problem of Memory

18 *"regularly employed under the same metrical conditions . . .":* Milman Parry, *The Making of Homeric Verse: The Collected Papers of Milman Parry,* ed. Adam Parry (Oxford: Oxford University Press, 1971), p. 272; also see his earlier definition, on p. 13.

18 *"groups of ideas regularly used in telling a tale . . .":* Albert B. Lord, *The Singer of Tales* (Cambridge, Mass.: Harvard University Press, 1964), p. 68.

19 *"those building blocks made up of rhythmic units . . .":* Eric A. Havelock, *Preface to Plato* (Cambridge, Mass.: Harvard University Press, 1963), p. 82.

20 *"The real and essential 'formula' . . .":* ibid.

20 *"formulaic state of mind":* ibid., p. 140.

21 *"a review and rearrangement of . . .":* ibid., p. 208.

21 *"separation of the knower from the known":* ibid., chap. 11, passim.

21 *"And so the [pages of Plato's Republic] are filled . . .":* ibid., p. 217.

22 *"We shall deliberately adopt the hypothesis . . .":* ibid., p. 66.

24 *"A 'map' is a visual concept . . .":* James Fentress and Chris Wickham, *Social Memory* (Oxford: Blackwell, 1992), p. 17.

25 *"Images can be transmitted socially . . .":* ibid., pp. 47–48.

27 *"Concerning social memory in particular . . .":* Paul Connerton, *How Societies Remember* (Cambridge: Cambridge University Press, 1989), p. 3.

27 *"Social memory is not stable as information . . .":* Fentress and Wickham, *Social Memory,* p. 59.

Chapter Two: Early Literacy and List Making

33 *"the picture of the voice":* for a standard account of the origins of writing, see I. J. Gelb, *A Study of Writing,* 2d ed. (Chicago: University of Chicago Press, 1963), who quotes Voltaire on p. 13.

35 *"time-factored activity":* the expression is from Alexander Marshack, *The Roots of Civilization* (New York: McGraw-Hill, 1972), p. 14.

36 *"token-iterative":* for the terms 'token' and 'token-iterative', and the terms 'emblem' and 'emblem slotting' (discussed below), see Roy Harris, *The Origin of Writing* (La Salle, Ill.: Open Court, 1986), pp. 131–46.

40 *"enclave of contrived speech":* Eric A. Havelock, *The Greek Concept of Justice* (Cambridge, Mass.: Harvard University Press, 1978), p. 30.

43 *"40 kagu-breads . . .":* Hans J. Nissen, Peter Damerow, and Robert K. Englund, *Archaic Bookkeeping: Early Writing and Techniques of Economic Administration in the Ancient Near East,* trans. Paul Larsen (Chicago: University of Chicago Press, 1993), p. 47.

45 *"If a man's chest-hair curls upwards . . .":* Jean Bottéro, *Mesopotamia: Writing, Reasoning, and the Gods,* trans. Zainab Bahrani and Marc van de Mieroop (Chicago: University of Chicago Press, 1992), p. 127.

46 *"trees, wooden objects, reeds . . .":* A Leo Oppenheim, *Ancient Mesopotamia: Portrait of a Dead Civilization* (Chicago: University of Chicago Press, 1964), p. 247.

47 *"a voluntary activity or occupation . . .":* Johan Huizinga, *Homo Ludens: A Study of the Play Element in Culture* (Boston: Beacon, 1955), p. 28.

47 *"Every group of names . . .":* Edward Chiera, *Sumerian Lexical Texts from the Temple School of Nippur* (Chicago: University of Chicago Press, 1929), p. 2.

47 *"polyphony":* the term is from Bottéro, *Mesopotamia,* p. 90.

48 *"Asari":* for this example, see ibid., pp. 88, 94–95.

49 *"logocentrism"* and *"reocentrism":* we here rely on Roy Harris's use of these terms, which differs from that of some recent postmodern thinkers. See Roy Harris, *The Language-Makers* (Ithaca, N.Y.: Cornell University Press, 1980), pp. 48–49.

49 *"stars and meteorites . . .":* Bottéro, *Mesopotamia,* p. 127.

50 *"If he urinates upwards . . .":* ibid., p. 116.

52 *"Etana, a shepherd, the one who . . .":* Thorkild Jacobsen, *The Sumerian King List* (Chicago: University of Chicago Press, 1939), p. 81.

53 *"When the kingship was lowered . . .":* ibid., p. 71.

Chapter Three: Alphabetic Literacy and the Science of Classification

65 *"principle of economy":* Gelb, *Study of Writing,* p. 69.

72 *"Be a friend to your friend . . .":* Hesiod, *Works and Days* 2.353–56; the translation is from *Hesiod,* trans. Richard Lattimore (Ann Arbor: University of Michigan Press, 1959).

73 *"seem to talk to you . . .":* Plato, *Phaedrus* 274e–275e; all translations of Plato are from *The Collected Dialogues of Plato,* ed. Edith Hamilton and Huntington Cairns (New York: Random House, 1961).

74 *"If things are many . . .":* Plato, *Parmenides* 127e.

74 *"likeness itself"* and *"unlikeness itself":* ibid., 129a.

74 *"unity"* and *"plurality":* ibid., 129c–d.

75 *"tumbling in a bottomless pit of nonsense":* ibid., 130c–e.

75 *"[The philosopher] discerns clearly . . .":* Plato, *Sophist* 253d–e.

76 *"Names, I maintain, are in no case stable . . .":* Plato, *Seventh Letter* 343b.

77 *"when, suddenly, like a blaze . . .":* ibid., 341d.

77 *"equivocally," "derivatively,"* and *"univocally":* Aristotle, *Categories* 1a1–15; all translations of Aristotle are from *The Complete Works of Aristotle,* ed. Jonathan Barnes, 2 vols. (Princeton: Princeton University Press, 1984).

78 *"language follows thought . . .":* Frederick Copleston, S. J., *A History of Philosophy,* rev. ed., 8 vols. (Garden City, N.Y.: Doubleday, 1962), vol. 1, part 2, p. 24.

79 *"what is said of it":* José Ortega y Gasset, *The Idea of Principle in Leibniz and the Evolution of Deductive Theory,* trans. Mildred Adams (New York: W. W. Norton, 1971), p. 58.

79 *"knowing"* and *"perceiving":* Marjorie Greene, *A Portrait of Aristotle* (Chicago: University of Chicago Press, 1963), p. 46.

82 *"substantial form":* for this term and the examples illustrating its meaning, see D. J. Allan, *The Philosophy of Aristotle,* 2d ed. (Oxford: Oxford University Press, 1970), pp. 83–85; and Jonathan Lear, *Aristotle: The Desire to Understand* (Cambridge: Cambridge University Press, 1988), pp. 273–93.

83 *"reasons why"* and *"causes":* Aristotle, *Posterior Analytics* 71b9–11. For the equivalency between the terms 'reasons why' and 'causes', see John Herman Randall Jr., *Aristotle* (New York: Columbia University Press, 1960), pp. 123–24.

Chapter Four: Printing and the Rupture of Classification

87 *"now more than half run out":* the phrase is from the Latin inscription commemorating Montaigne's retirement, painted on a wall in his tower/library retreat. The translation is from Donald M. Frame, *Montaigne: A Biography* (New York: Harcourt, Brace, and World, 1965), p. 115, with the original Latin in an endnote.

88 *"chimeras and fantastic monsters":* Michel de Montaigne, *The Complete Works of Montaigne,* trans. Donald M. Frame (Stanford: Stanford University Press, 1958), p. 21.

88 *"A man born in 1453 . . .":* quoted in Elizabeth L. Eisenstein, *The Printing Press as an Agent of Change,* 1 vol. ed. (Cambridge: Cambridge University Press, 1979), p. 45.

93 *"mass of words":* quoted in M. T. Clanchy, *From Memory to Written Record: England, 1066–1307,* 2d ed. (Oxford: Blackwell, 1993), p. 107.

98 *"The object of the will . . .":* Francesco Petrarca, "On His Own Ignorance and That of Many Others," in *The Renaissance Philosophy of Man,* ed. Ernst Cassirer, Paul Oskar Kristeller, and John Herman Randall Jr. (Chicago: University of Chicago Press, 1948), p. 105.

99 *"Everyone who has become . . .":* ibid., p. 104.

105 *"rolle"* and *"registre":* Michel de Montaigne, *Essais,* in *Oeuvres complètes de Montaigne,* ed. Albert Thibaudet and Maurice Rat, Bibliothèque de la Pléiade (Paris: Gallimard, 1962), pp. 34, 648.

105 *"I have no more made my book . . .":* Montaigne, *Complete Works,* p. 504.

105 *"strange semi-ruinous mass":* J. G. A. Pocock, *The Ancient Constitution and the Feudal Law* (Cambridge: Cambridge University Press, 1957), p. 11.

106 *"Now trust in your philosophy . . .":* Montaigne, *Complete Works,* p. 383.
106 *"an instrument of lead and of wax . . .":* ibid., p. 425.
106 *"On an empty stomach . . .":* ibid.
107 *"What do I know?":* ibid., p. 393.
107 *"We are Christians . . .":* ibid., p. 325.
108 *"clear and distinct ideas":* René Descartes, *Discourse on the Method,* in *The Philosophical Works of Descartes,* trans. and ed. Elizabeth S. Haldane and G. R. T. Ross, 2 vols. (Cambridge: Cambridge University Press, 1970), 1:92 (hereafter cited as *PWD*).
110 *"When one understands the causes . . .":* quoted in Zachary Sayre Schiffman, *On the Threshold of Modernity: Relativism in the French Renaissance* (Baltimore: Johns Hopkins University Press, 1991), p. 112.

110 *"forming images dependent upon one another":* ibid.
111 *"uses up too much paper":* ibid.

Chapter Five: Numeracy, Analysis, and the Reintegration of Knowledge

112 *"the modern world . . .":* Philip J. Davis and Reuben Hersh, *Descartes' Dream: The World According to Mathematics* (Boston: Houghton Mifflin, 1986), p. 3.
112 *"What path shall I follow . . .":* quoted in Jacques Maritain, *The Dream of Descartes,* trans. Mabelle L. Andison (New York: Philosophical Library, 1944), p. 14.
113 *"foundations of the marvelous science":* quoted in Stephen Gaukroger, *Descartes: An Intellectual Biography* (Oxford: Oxford University Press, 1995), p. 105.
113 *"inspiration and spiritual father":* quoted in S. V. Keeling, *Descartes* (London: Oxford University Press, 1968), p. 10.
113 *"of rightly conducting . . .":* Descartes, *Discourse on the Method,* in *PWD,* 1:79.
113 *mathesis universalis:* Descartes, *Rules for the Direction of the Mind,* in *PWD,* I:14.
113 *"order and disposition . . .":* ibid.
113 *"philosophy and wisdom . . .* " and *"all the various sciences . . .":* quoted in Maritain, *Dream,* p. 14.
114 *"And new philosophy . . .":* John Donne, "An Anatomy of the World: The First Anniversary," in A. L. Clements, ed., *John Donne's Poetry* (New York: W. W. Norton, 1966), p. 73.
114 *"imagination mathematical":* quoted in Marie Boas, *The Scientific Renaissance: 1450–1630* (New York: Harper & Row, 1962), p. 197.
115 *"about in a dark labyrinth":* Galileo Galilei, "The Assayer," in Stillman Drake, ed. and trans., *Discoveries and Opinions of Galileo* (Garden City, N.Y.: Doubleday, 1957), p. 238.
118 *"progresses in the sense . . .":* Tobias Dantzig, *Number: The Language of Science,* 4th ed. (Garden City, N.Y.: Doubleday, 1954), p. 8.
118 *"collective unit":* John D. Barrow, *Pi in the Sky: Counting, Thinking, and Being* (Oxford: Clarendon Press, 1992), p. 51.
120 *"common boundary":* Aristotle, *Categories* 4b20–5a17.
120 *"infinitely divisible":* Aristotle, *Physics* 3.200b19.
120 *"together at a common boundary":* Aristotle, *Categories* 4b20–5a17.

120 *"Arithmetic is about . . .":* Aristotle, *Posterior Analytics* 1.76b4 and 1.75a38.

121 *"the most successful intellectual . . .":* Barrow, *Pi in the Sky,* p. 92.

123 *"relation-mathematics"* and *"thing-mathematics":* Jagjit Singh, *Great Ideas of Modern Mathematics: Their Nature and Use* (New York: Dover, 1959), p. 8.

123 *"the calculation of reduction . . .":* D. J. Struik, ed., *A Source Book in Mathematics, 1200–1800* (Princeton: Princeton University Press, 1986), p. 56.

124 *"rhetorical algebra"* and *"syncopated algebra":* Tobias Dantzig, *Number: the Language of Science* (Garden City, N.Y.: Doubleday Anchor Books, 1954), p. 79.

124 *"general letter algebra":* François Viète, *Introduction to the Analytic Art,* in Struik, ed., *Source Book,* p. 74.

124 *"threefold analytic art," "logistica speciosa,"* and *"logistica numerosa":* ibid., pp. 75, 78.

124 *"law of homogeneity, . . .":* ibid., p. 76.

124 *"numerable"* and *"measurable":* Aristotle, *Metaphysics* 1020a7–14.

125 *"scalar magnitudes," "by their own nature . . . ," "side or root,"* and *"squared square":* Viète, *Analytic Art,* in Struik, ed., *Source Book,* p. 77.

125 *"(1) Length and breadth; . . .":* ibid.

125 *"by means of species . . .":* ibid., p. 78.

125 *"Let the given number . . .":* ibid., p. 81.

126 *"where he [Viète] left off":* "Descartes à Mersenne, décembre 1637," in *Oeuvres philosophiques,* 3 vols. (Paris: Editions Garnier Frères, 1963), 1:821–22. All translations from this edition are ours.

127 *"general science," "discovery of an order," "facts," "interval, greater, less, or equal," "chief secret of the method," "correlative connection and natural order":* Descartes, *Rules,* in *PWD,* 1:13, 64, 15, 16.

127 *"pure mathematics . . . proportions or relations of things":* ibid., 1:17.

127 *"detached and unencumbered way"* and *"the abstract formulation is . . .":* ibid., 1:67–69.

127 *"ontological genus . . .":* ibid., 1:15.

127 *"our nephews will find nothing . . .":* "Descartes à Mersenne, décembre 1637," *Oeuvres philosophiques,* 1:822.

128 *"whole nomenclature . . . ," "must be abandoned,"* and *"cube or biquadratic . . .":* Descartes, *Rules,* in *PWD,* 1:68. We have used our own translation here; see *Oeuvres philosophiques,* 1:188.

128 *"at which to begin . . ."* and *"all points of those curves . . .":* René Descartes, *The Geometry of René Descartes,* trans. David Eugene Smith and Marcia L. Latham (New York: Dover, 1954), pp. 48–51.

133 *"the very core . . .":* Edna E. Kramer, *The Nature and Growth of Modern Mathematics* (Princeton: Princeton University Press, 1981), p. 137.

133 *"Information is a name . . .":* Norbert Wiener, *The Human Use of Human Beings: Cybernetics and Society,* afterword by Walter A. Rosenblith (New York: Avon Books, 1967), pp. 26–27.

133 *"pure and simple natures,":* Descartes, *Rules,* in *PWD,* 1:16.

134 *"resolutive-compositive,":* see E. J. Dijksterhuis, *The Mechanization of the World Picture,* trans. C. Dikshoorn (Oxford: Oxford University Press, 1961), p. 339.

134 *"compounded motion"* and *"moveables":* Galileo Galilei, *Two New Sciences,* trans. and intro. Stillman Drake (Madison: University of Wisconsin Press, 1974), pp. 216–22.

136 *"All teaching . . .":* Aristotle, *Posterior Analytics* 71a1.

136 *"primitive notions"* and *"models":* René Descartes, *Philosophical Letters,* trans. and ed. Anthony Kenny (Oxford: Clarendon Press, 1970), p. 138.

136 *"philosophy as a whole . . . whose roots . . .":* Descartes, *Principles of Philosophy,* in *PWD,* 1:211.

137 *"systematic spirit":* Jean Le Rond d'Alembert, *Preliminary Discourse to the Encyclopedia of Diderot,* trans. Richard N. Schwab (Indianapolis: Bobbs-Merrill, 1963), p. 22.

Chapter Six: The Analytical World Map

146 *"written this huge book . . .":* the story is retold in Eric Temple Bell, *Men of Mathematics* (New York: Simon & Schuster, 1986), p. 81.

146 *"Ah, but that is a fine hypothesis . . .":* ibid.

147 *"God made the integers, . . .":* quoted in Stuart Hollingdale, *Makers of Mathematics* (New York: Penguin, 1989), p. 15.

147 *"ghosts of departed quantities":* quoted in ibid., p. 305.

147 *"the subject in which we never know . . .":* quoted in Carl B. Boyer, *The History of the Calculus and Its Conceptual Development* (New York: Dover, 1949), p. 3.

148 *"horrible mass of books . . .": Leibniz: Selections,* ed. Philip P. Wiener (New York: Charles Scribner's Sons, 1951), p. 29.

148 *"enchainment of knowledge":* Denis Diderot and Jean Le Rond d'Alembert, eds., *Encyclopédie, ou dictionnaire raisonné des sciences, des arts et des métiers, par un société de gens des lettres* (Paris, 1751–65), 5:635. All translations from the *Encyclopédie* are ours, unless otherwise noted.

148 *"detailed system of human knowledge":* d'Alembert, *Preliminary Discourse,* pp. 143–57.

148 *"spirit of the system":* ibid., p. 23.

149 *"world map," "great regions,"* and *"individual maps": Encyclopédie,* 5:641A.

149 *"Systematic Chart"* and *"order and connection":* d'Alembert, *Preliminary Discourse,* pp. 4, 57.

150 *"fluxions":* for this and other early calculus terms, see Boyer, *History of Calculus,* pp. 189–200.

152 *"phantom":* d'Alembert, *Preliminary Discourse,* p. 19.

157 *"the art of exactly numbering . . .":* François Marie Arouet de Voltaire, *Philosophical Letters,* trans. Ernest Dilworth (Indianapolis: Bobbs-Merrill, 1961), p. 79.

159 *"ultimate ratio . . .":* see Boyer, *History of the Calculus,* chap. 5, passim, for the various terms.

159 *"he who can digest . . .":* quoted in Morris Kline, *Mathematical Thought from Ancient to Modern Times* (New York: Oxford University Press, 1972), p. 428.

160 *"language whose signs . . .":* C. A. van Peursen, *Leibniz: A Guide to His Philosophy,* trans. Hubert Hoskins (New York: Dutton, 1970), pp. 32–36; Boyer, *History of Calculus,* p. 209; Wiener, ed., *Leibniz: Selections,* p. 18.

161 *"edge of objectivity"*: Charles Coulston Gillispie, *The Edge of Objectivity: An Essay in the History of Scientific Ideas* (Princeton: Princeton University Press, 1960).
162 *"explain the general system . . ."*: *Encyclopédie*, 5:635.
162 *"encyclopedic arrangement"*: d'Alembert, *Preliminary Discourse*, p. 47.
162 *"first principles, . . ."*: *Encyclopédie*, 5:641A.
163 *"method of solving . . ."*: ibid., 1:400.
163 *"an abstract relation . . ."*: ibid., 11:202.
163 *"operation"*: ibid., 5:494.
163 *"substantial forms," "divest matter of almost . . . ," "only its phantom,"*: d'Alembert, *Preliminary Discourse*, pp. 6, 19.

163 *"there is no [intrinsic] connection . . ."* and *"Only a kind of instinct . . ."*: ibid., p. 9.
164 *"Thus the number 3 expresses . . ."*: *Encyclopédie*, 1:675.
164 *"everything . . . susceptible to . . ."*: d'Alembert, *Preliminary Discourse*, p. 20.
164 *"certain rules relative to"* and *"abbreviated manners"*: *Encyclopédie*, 1:675–76.
165 *"particular arithmetics," "an art of making . . . ,"* and *"numbers by the different . . ."*: ibid.
165 *"general"* and *"same point of view"*: ibid., 7:550.
166 *"a large number of . . ."*: ibid.
166 *"general result . . . ," "easy method,"* and *"to reduce . . ."*: ibid., 7:184–85.
168 *"to learn entire . . ."*: ibid., 1:400.
168 *"infinitesimal"* and *"infinitely small quantity"*: ibid., 8:805.
168 *"method of calculating . . ."*: ibid., 1:259.
168 *"the more one reduces . . ."*: d'Alembert, *Preliminary Discourse*, pp. 22, 152.
168 *"mixed mathematics"*: *Encyclopédie*, 1:854.
170 *"that belong in a certain sense . . ."*: ibid., 6:301.
170 *"If the perhaps infinite sum . . ."*: quoted in Lester Crocker, *Diderot's Chaotic Order: Approach to Synthesis* (Princeton: Princeton University Press, 1974), p. 44.
171 *"most natural and rigorous . . . the elements of all the sciences . . ."*: *Encyclopédie*, 5:491.
171 *"The universe . . ."*: d'Alembert, *Preliminary Discourse*, p. 29.

Chapter Seven: Analysis Uprooted

175 *"Sir Alphabet Function"*: quoted in Joel Shurkin, *Engines of the Mind: A History of the Computer* (New York: W. W. Norton, 1984), p. 35. Babbage referred to himself as:

> Sir Alphabet Function, a knight much renowned,
> Who had gained little credit on classical ground,
> Set out through the world his fortune to try,
> With nought to his pate but his x, v and y.

175 *"1st. the store . . ."*: Philip Morrison and Emily Morrison, eds., *Charles Babbage, On the Principles and Development of the Calculator, and Other Seminal Writings by Charles Babbage and Others* (New York: Dover, 1961), p. 55.

177 *"science of order"* and *"postulational-deductive science":* John Passmore, *A Hundred Years of Philosophy* (Middlesex, England: Penguin, 1970), pp. 145–46; Kramer, *Nature and Growth of Modern Mathematics,* p. 79.

178 *"laws of the symbol . . .":* George Boole, *An Investigation of The Laws of Thought, On which are Founded the Mathematical Theories of Logic and Probabilities* (New York: Dover, n.d.), pp. 44–45.

180 *"law"* and *"a differential equation":* Henri Poincaré, *The Value of Science,* trans. George Bruce Halsted (New York: Dover, 1958), p. 93.

185 *"characteristic function":* quoted in Thomas L. Hankins, *Sir William Rowan Hamilton* (Baltimore: Johns Hopkins University Press, 1980), p. 78.

188 *"arithmetical"* and *"symbolical algebra":* quoted in Helena M. Pycior, "Augustus De Morgan's Algebraic Work: The Three Stages," *Isis* 74 (1983): pp. 218, 211.

189 *"tied to":* quoted in Passmore, *A Hundred Years of Philosophy,* p. 122.

189 *"For the first time . . .":* ibid., p. 123.

189 *"fundamental laws . . .":* Boole, *Laws of Thought,* p. 1.

189 *"language and number"* and *"instrumental aids . . .":* ibid., p. 2.

190 *"the foundation of . . .":* ibid., p. 5.

190 *"fundamental laws . . .":* ibid., p. 3.

190 *"This is the work . . .":* quoted in Anthony Hyman, *Charles Babbage: Pioneer of the Computer* (Princeton: Princeton University Press, 1983), p. 244.

190 *"certain general principles . . ."* and *"we never . . .":* Boole, *Laws of Thought,* p. 6.

190 *"ultimate laws of thought":* ibid., p. 11.

191 *"abstraction," "mental operation,"* and *"process":* ibid., p. 37.

191 *"first step . . . conception of . . .":* ibid., p. 43.

191 *"selection" . . . "classes of objects":* ibid.

191 *"arbitrary mark" . . . "all beings":* ibid., pp. 26–28.

191 *"substantive verb . . .":* ibid., p. 34.

192 *"forming the aggregate . . .":* ibid., p. 32.

194 *"a Calculus"* and *"acquaintance with . . .":* ibid., p. 1 and preface, n.p.

194 *"index law"* and *"idempotent":* quoted in C. I. Lewis, *A Survey of Symbolic Logic* (New York: Dover, 1960), p. 54; Kramer, *Nature and Growth of Modern Mathematics,* p. 107.

194 *"the proposition X . . .":* Boole, *Laws of Thought,* p. 169.

194 *"two-valued algebra":* William Kneale and Martha Kneale, *The Development of Logic* (Oxford: Oxford University Press, 1962), pp. 413–20.

198 *"the smallest wheels . . .":* Heinz R. Pagels, *The Cosmic Code: Quantum Physics as the Language of Nature* (New York: Bantam Books, 1982), p. 47.

198 *"class in which . . .":* Boole, *Laws of Thought,* p. 48.

Chapter Eight: The Realm of Pure Technique

202 *"ether of our times":* Robert W. Lucky, *Silicon Dreams: Information, Man, and Machine* (New York: St. Martin's Press, 1989), p. 1.

210 *"digital logical circuitry":* Arnio Penzias, *Ideas and Information: Managing in a High-Tech World* (New York: Simon & Schuster, 1989), p. 100.

213 *"heretic scientist . . .":* Andrew Hodges, *Alan Turing: The Enigma* (New York: Simon & Schuster, 1983), p. 523.

214 *"mechanical set of rules," "automatic,"* and *"the behavior of . . ."*: quoted in ibid., p. 105.

214 *"state of mind," "simple,"* and *"so elementary . . ."*: quoted in ibid.

215 *"The state of progress . . ."*: quoted in ibid., p. 107.

215 *"no essential distinction . . ."*: quoted in ibid., p. 102.

215 *"as simple as possible"*: quoted in ibid., p. 320.

217 *"binary coding of . . ."*: Roger Penrose, *The Emperor's New Mind: Concerning Computers, Minds and the Laws of Physics* (New York: Penguin, 1991), p. 42.

219 *"paragon of science"*: Steve J. Heims, *John von Neumann and Norbert Wiener: From Mathematics to the Technologies of Life and Death* (Cambridge, Mass.: MIT Press, 1980), p. 358.

219 *"von Neumann architecture"*: Herman H. Goldstine, *The Computer from Pascal to von Neumann* (Princeton: Princeton University Press, 1972), p. 202.

220 *"flow diagram"*: For this and other terms cited in our discussion of von Neumann, see John von Neumann, *The Computer and the Brain* (New Haven: Yale University Press, 1958), passim, but especially part 1; *Theory of Self-reproducing Automata,* ed. and completed by Arthur W. Burks (Urbana: University of Illinois Press, 1966); and Arthur W. Burks, Herman H. Goldstine, and John von Neumann, "Preliminary Discussion of the Logical Design of an Electronic Computing Instrument," in John Diebold, ed., *The World of the Computer* (New York: Random House, 1973), 44–82.

224 *"hierarchy"*: Lucky, *Silicon Dreams,* p. 19.

224 *"entirely of the radiations . . ."*: Bertrand Russell, *An Outline of Philosophy* (London: Routledge, 1993), pp. 124–26.

226 *"computopia . . ."*: Yoneji Masuda, "Computopia," in Tom Forester, ed., *The Information Technology Revolution* (Cambridge, Mass.: MIT Press, 1985), p. 624.

Chapter Nine: Information Play

235 *"voluntary activity . . ."*: Huizinga, *Homo Ludens,* p. 28.

237 *"atoms"* and *"bits"*: Nicholas Negroponte, *Being Digital* (New York: Vintage Books, 1996), pp. 11–17 and passim.

238 *"formless data"*: ibid., p. 61.

238 *"culture arises . . ."*: Huizinga, *Homo Ludens,* p. 46.

239 *"proceeds in the shape . . ."*: ibid.

240 *"existing jobs . . ."*: Penzias, *Ideas and Information,* p. 25.

241 *"hypertexts"*: Jay David Bolter, *Writing Space: The Computer, Hypertext, and the History of Writing* (Hillsdale, N.J.: Lawrence Erlbaum, 1991), pp. 92–97 and passim.

241 *"Nature to be commanded . . ."*: Francis Bacon, *Novum Organum,* in *The Works of Francis Bacon,* 2 vols. (New York: Hurd and Houghton, 1878), p. 68.

242 *"he has five brains . . ."*: quoted in David Berreby, "The Man Who Knows Everything: Murray Gell-Mann," *New York Times Magazine,* May 8, 1994, p. 26.

242 *"from the point of view . . ."* and *"complex adaptive systems"*: Murray Gell-Mann, interview in John Brockman, *The Third Culture* (New York: Simon & Schuster, 1996), pp. 317, 319.

245 *"quantitative measure"* and *"simple order . . ."*: Heinz R. Pagels, *The Dreams of*

Reason: The Computer and the Rise of the Sciences of Complexity* (New York: Simon & Schuster, 1988), p. 54;

245 *"orderly enough . . .":* Stuart Kauffman, *At Home in the Universe: The Search for Laws of Self-organization and Complexity* (New York: Oxford University Press, 1995), p. 87.

247 *"cellular automata . . .":* Pagels, *Dreams of Reason,* pp. 97–99.

247 *"artificial life," "central cells,"* and *"1. The cell will be . . .":* See John L. Casti, *Complexification: Explaining a Paradoxical World through the Science of Surprise* (New York: Harper Collins, 1994), p. 223.

250 *"edge of chaos":* see, for example, Kauffman, *At Home,* p. 26.

250 *"heretical"* and *"renegade":* ibid., pp. 99, 26.

250 *"self-organization," "order for free,"* and *"magic of template replication":* ibid., pp. 26, 71, 47.

250 *"chance caught on the wing":* quoted in ibid., p. 71.

251 *"computer-simulation movies":* ibid., p. 64.

251 *"spontaneous order":* ibid., p. 71.

254 *"sudden and stunning":* ibid., p. 83.

255 *"self-reproducing chemical networks . . .":* ibid., p. 73.

256 *"arrow of time":* see, for example, Peter Coveney and Roger Highfield, *The Arrow of Time: A Voyage through Science to Solve Time's Greatest Mystery* (New York: Fawcett Columbine, 1991), passim.

258 *"coarse graining"* and *"Any definition . . .":* Murray Gell-Mann, *The Quark and the Jaguar: Adventures in the Simple and the Complex* (New York: W. H. Freeman, 1994), pp. 33, 35.

259 *"freedom"* and *"creates order,* is *order":* Huizinga, *Homo Ludens,* pp. 8, 15.

Conclusion: The Two Cultures and the Arrow of Time

261 *"inscrutable," "gods of Homer," "physical objects . . . ,"* and *"more efficacious . . .":* Willard Van Orman Quine, "Two Dogmas of Empiricism," in *From a Logical Point of View* (New York: Harper & Row, 1961), p. 44.

262 *"cultural DNA":* Gell-Mann, *The Quark and the Jaguar,* p. 292.

262 *"fresh support . . .":* Kauffman, *At Home,* p. 5.

263 *"emerging global civilization . . .":* ibid., pp. 299–300.

268 *"dreams of a final theory":* Steven Weinberg, *Dreams of a Final Theory* (New York: Vintage Books, 1994).

BIBLIOGRAPHICAL ESSAY

Chapter One: Orality and the Problem of Memory

Some scholars maintain that orality and literacy define different forms of consciousness and culture. This view owes much to the almost simultaneous appearance of three works: (1) Jack Goody and Ian Watt's seminal article "The Consequences of Literacy," first published in *Comparative Studies in Society and History* (1963) and reprinted in *Literacy in Traditional Societies,* ed. Jack Goody (Cambridge: Cambridge University Press, 1968); (2) Eric A. Havelock's *Preface to Plato* (Cambridge, Mass.: Harvard University Press, 1963); and (3) Marshall McLuhan's *The Gutenberg Galaxy* (Toronto: University of Toronto Press, 1962). Whereas McLuhan has fallen into scholarly disrepute, Goody and Havelock (along with Walter Ong, cited below) remain the mainstays of what has become known as the "great divide" theory of literacy.

Goody further elaborated the anthropological significance of this divide in a number of works, most notably *The Domestication of the Savage Mind* (Cambridge: Cambridge University Press, 1977), *The Logic of Writing and the Organization of Society* (Cambridge: Cambridge University Press, 1986), and *The Interface between the Written and the Oral* (Cambridge: Cambridge University Press, 1987). Havelock, too, continued to expound his views on the effects of alphabetic literacy in Greece, especially in *The Greek Conception of Justice* (Cambridge, Mass.: Harvard University Press, 1978), *The Literate Revolution in Greece and Its Cultural Consequences* (New Haven: Yale University Press, 1982), and *The Muse Learns to Write* (New Haven: Yale University Press, 1986).

The prolific and formidable Walter J. Ong, S.J., dubbed "the thinking man's McLuhan" (indeed, he may have even inspired McLuhan's best work), has also been a strong proponent of the great divide theory. See his major works: *Ramus: Method and the Decay of Dialogue* (Cambridge, Mass.: Harvard University Press, 1958), *The Presence of the Word* (New Haven: Yale University Press, 1967), *Rhetoric, Romance, and Technology* (Ithaca, N.Y.: Cornell University Press, 1971), and *Orality and Literacy* (London: Methuen, 1982).

The great divide theory initially garnered widespread acceptance, largely because the distinction between oral and literate cultures offered a more value-free alternative to the earlier one between "primitive" and "civilized" peoples. Lately, though, the theory has become an intellectual battleground, with opponents maintaining that it reduces complex historical and cultural developments to mere matters of technology. The position of the opponents is epitomized in Brian V. Street, *Literacy in Theory and Practice* (Cambridge: Cambridge University Press, 1984). Also see G. E. R. Lloyd, *The Revolutions of Wisdom* (Berkeley: University of California Press, 1987); Michael Heim, *Electric Language: A Philosophical Study of Word Processing* (New Haven: Yale University Press, 1987); Harvey J. Graff, *The Labyrinths of Literacy: Reflections on Literacy Past and Present* (London: Falmer Press, 1987) and *The Legacies of Literacy* (Bloomington: Indiana University Press, 1987); Ruth Finnegan, *Literacy and Orality* (Oxford: Basil Blackwell, 1988); William V. Harris, *Ancient Literacy* (Cambridge, Mass.: Harvard University Press, 1989); Rosalind Thomas, *Oral Tradition and Written Record in Classical Athens* (Cambridge: Cambridge University Press, 1989); and John Halverson, "Havelock on Greek Orality and Literacy," *Journal of the History of Ideas* 53 (1992): 148–63. Ironically, a volume of essays honoring Havelock, *Language and Thought in Early Greek Philosophy,* ed. Kevin Robb (La Salle, Ill.: Monist Library of Philosophy, 1983), also contains a number of pieces critical of his views.

It is interesting to compare the above arguments about the effects of literacy—asserted both pro and con by historians, anthropologists, classicists, and literary critics—with the more restrained observations of a linguist. See Josef Vachek's *Written Language* (The Hague: Mouton, 1973) and *Written Language Revisited* (Amsterdam: John Benjamin, 1989), which take a functional approach to the differences between spoken, written, and printed language.

The intellectual pedigree of the great divide theory ultimately extends back to the pioneering work of Milman Parry and Albert B. Lord, who explore the formulaic nature of the Homeric epics, extending it to oral poetry in general. See Parry's collected works, *The Making of Homeric Verse: The Collected Papers of Milman Parry,* ed. Adam Parry (Oxford: Oxford University Press, 1971) and Lord's *The Singer of Tales* (Cambridge, Mass.: Harvard University Press, 1964). Subsequent scholars have somewhat modified Parry's interpretation of the formulaic tradition in Homer; see, among others, J. B. Hainsworth, *The Flexibility of the Homeric Formula* (Oxford: Oxford University Press, 1968). For a recent attack on Parry's notion of the economy of Homeric formulas, see David M. Shive, *Naming Achilles* (Oxford: Oxford University Press, 1987). Some reviewers have questioned whether Shive really understands Parry's notion of economy; see, for example, the responses to Hugh Lloyd-Jones's article in *The New York Review of Books,* cited below. The standard survey of the field of oral poetry remains Ruth Finnegan's *Oral Poetry* (Cambridge: Cambridge University Press, 1977), which is at pains to go beyond Parry's and Lord's narrow view that all oral poetry is composed in performance. In addition, see Benjamin A. Stolz and Richard S. Shannon, eds., *Oral Literature and the Formula* (Ann Arbor: Center for the Coordination of Ancient and Modern Studies, 1976). The field of ethnomusicology can also enrich our understanding of oral poetry (which is, after all, song); see, for example, Paul F. Berliner's observations about improvisation and precomposition in his *Thinking in Jazz: The Infinite Art of Improvisation* (Chicago: University of Chicago Press, 1994).

The oral interpretation of Homer is as hotly debated as the great divide theory, in a body of scholarship too extensive to cite at any length. Suffice it to say that the oral interpretation is nicely represented by G. S. Kirk's *The Songs of Homer* (Cambridge: Cambridge University Press, 1962). Kirk addresses some of the objections to this interpretation in his *Homer and the Oral Tradition* (Cambridge: Cambridge University Press, 1976). The view of Homer as a literate poet is summarized by Hugh Lloyd-Jones in "Keeping Up with Homer," *The New York Review of Books* 39, no. 5 (1992): 52–57; also see the letters in response to this article, in *The New York Review of Books* 39, no. 9 (1992): 51–52, and 39, no. 12 (1992): 57–58.

For the role of memory in oral cultures, and especially for the activity of commemoration, see the indispensable survey by James Fentress and Chris Wickham, *Social Memory* (Oxford: Blackwell, 1992), to which we are greatly indebted. Also of general interest is Paul Connerton, *How Societies Remember* (Cambridge: Cambridge University Press, 1989).

Chapter Two: Early Literacy and List Making

Our interpretation of the origin of writing is based chiefly on the work of the Oxford linguist Roy Harris, *The Origin of Writing* (La Salle, Ill.: Open Court, 1986). He rejects the "tyranny of the alphabet" in prior accounts of the origins of writing and emphasizes the need to distinguish more clearly between picture drawing and writing. Also see Harris's *The Language-Makers* (Ithaca, N.Y.: Cornell University Press, 1980), which describes how different writing systems lead to different views of language and its relation to the things of the world. For a standard account of the origins of writing, see I. J. Gelb, *A Study of Writing,* 2d ed. (Chicago: University of Chicago Press, 1963). In *A History of Writing* (New York: Charles Scribner's Sons, 1984), Albertine Gaur views the invention more from the late-twentieth-century perspective of information storage rather than language representation.

Roy Harris argues in *Origins of Writing* that the development of accounting led to "emblem slotting," the innovation that divided picture drawing from writing. This view derives in part from Alexander Marshack's innovative work *The Roots of Civilization* (New York: McGraw-Hill, 1972), which describes the earliest evidence of tallying, a form of record keeping that led to counting. Further developments in record keeping are detailed in Denise Schmandt-Besserat, *Before Writing,* 2 vols. (Austin: University of Texas Press, 1992), which by and large confirms Harris's hypothesis about the origins of writing, despite differences of approach and terminology. Also see Schmandt-Besserat's influential yet controversial article "The Earliest Precursor of Writing," *Scientific American* 238 (1978): 50–59, and (among others) I. J. Gelb's response to it, "Principles of Writing Systems within the Frame of Visual Communication," in Paul A. Kolers, Merald E. Wrolstad, and Herman Bouma, eds., *Processing Visible Language 2* (New York: Plenum Press, 1979). In addition to these works, the collaborative effort by Hans J. Nissen, Peter Damerow, and Robert K. Englund, *Archaic Bookkeeping: Early Writing and Techniques of Economic Administration in the Ancient Near East,* trans. Paul Larsen (Chicago: University of Chicago Press, 1993) proffers an indispensable analysis of Mesopotamian accounting practices and their relation to writing. See too the relevant articles on the Sumerian language, writing, and literature i. volume 4 of *Civilizations of the Ancient Near East,* ed. Jack M. Sasson, 4 vols. (New York: Charles Scribner's Sons, 1995).

Jean Bottéro's *Mesopotamia: Writing, Reasoning, and the Gods,* trans. Zainab Bahrani and Marc van de Mieroop (Chicago: University of Chicago Press, 1992) provides a penetrating account of the intellectual effects of Mesopotamian writing, especially in regard to list making. On the latter topic, see in general Goody's *Domestication of the Savage Mind* (cited above) and A. Leo Oppenheim, *Ancient Mesopotamia: Portrait of a Dead Civilization* (Chicago: University of Chicago Press, 1964), which describes the exhaustive nature of Mesopotamian list making. This aspect is also alluded to in the brief introduction to Edward Chiera, *Sumerian Lexical Texts for the Temple School of Nippur* (Chicago: University of Chicago Press, 1929). For a translation and analysis of an actual list, see Thorkild Jacobsen, *The Sumerian King List* (Chicago: University of Chicago Press, 1939). Henri Frankfort's *Kingship and the Gods* (Chicago: University of Chicago Press, 1948) describes the theological basis for the Mesopotamian idea of kingship; and Seton Lloyd's *The Archaeology of Mesopotamia* (London: Thames and Hudson, 1978) helps contextualize the King List, providing an introduction to the anatomy and chronology of Mesopotamian civilization.

Chapter Three: Alphabetic Literacy and the Science of Classification

Our view of the effects of alphabetic literacy was inspired by the conjunction of two passing comments: (1) a statement in the first chapter of Fentress and Wickham's above-cited *Social Memory* that the distinction between words and things became apparent only as writing increasingly modeled itself on language, and (2) Roy Harris's comment in the above-cited *Language-Makers* that Platonic philosophy serves to clarify the relation between words and things. Eric Havelock's essay "Thoughtful Hesiod," in his above-cited *Literate Revolution in Greece,* also influenced the formation of our ideas, by showing how the literate Hesiod exemplifies an early step toward the classifications of Plato.

John F. Healey, *The Early Alphabet* (Berkeley: University of California Press, 1990) provides a good, concise account of the origins of the alphabet. In the interests of brevity, Healey has necessarily summarized and simplified many complex arguments and debates. For more detailed analyses, see Gelb, *Origins of Writing* (cited above); David Diringer, *The Alphabet: A Key to the History of Mankind,* 3d ed., 2 vols. (New York: Funk & Wagnalls, 1968); and Barry B. Powell, *Homer and the Origin of the Greek Alphabet* (Cambridge: Cambridge University Press, 1991). Powell's arguments are speculative yet forceful, inferring about as much as one reasonably can from the scant evidence of early Greek alphabetic writing. For further observations regarding the Semitic sources of the Greek alphabet, see G. R. Driver, *Semitic Writing,* rev. ed. (Oxford: Oxford University Press, 1976) and Kevin Robb, "Poetic Sources of the Greek Alphabet: Rhythm and the Abecedarium from Phonecian to Greek," in *Communication Arts in the Ancient World,* ed. Eric A. Havelock and Jackson P. Hershbell (New York: Hastings House, 1978).

M. L. West, *The Hesiodic Catalogue of Women* (Oxford: Oxford University Press, 1985) provides a good introduction not only to the fragmentary *Catalogue of Women* but also to Greek genealogical poetry in general. Books on Platonic and Aristotelian philosophy are legion. Terence Irwin provides a concise introduction to Plato and Aristotle, placing them in their historical and philosophical context, in his *Classical Thought* (Oxford: Oxford University Press, 1989). For more detailed, traditional accounts of these figures, see the appropriate volumes of Frederick Copleston, S.J., *A History of Philosophy,* rev. ed., 9 vols. (Garden City, N.Y.: Doubleday, 1962), which we cite as but one of many standard surveys.

Older accounts of Plato should be supplemented with the pathbreaking work of Gregory Vlastos, especially in his collections of essays *Platonic Studies* (Princeton: Princeton University Press, 1973) and *Studies in Greek Philosophy: Socrates, Plato, and Their Tradition,* ed. Daniel W. Graham (Princeton: Princeton University Press, 1995). The latter volume, though somewhat technical, presents a stunning analysis of the *Parmenides.*

Terence Irwin's compendious *Aristotle's First Principles* (Oxford: Oxford University Press, 1988) furnishes the starting point for any investigation of Aristotle's philosophical method. Other useful works are Jonathan Lear, *Aristotle: The Desire to Understand* (Cambridge: Cambridge University Press, 1988); D. J. Allan, *The Philosophy of Aristotle,* 2d ed. (Oxford: Oxford University Press, 1970); G. E. R. Lloyd, *Aristotle: The Growth and Structure of His Thought* (Cambridge: Cambridge University Press, 1968); Marjorie Greene, *A Portrait of Aristotle* (Chicago: University of Chicago Press, 1963); and John Herman Randall Jr., *Aristotle* (New York: Columbia University Press, 1960). Our analysis of the tradition of Aristotelianism that dominated Western thought for millennia is indebted to two classic works: Ernst Cassirer, *Substance and Function, and Einstein's Theory of Relativity,* trans. William and Marie Swabey (New York: Dover, 1923), which is noteworthy for its concision; and José Ortega y Gasset, *The Idea of Principle in Leibniz and the Evolution of Deductive Theory,* trans. Mildred Adams (New York: W. W. Norton, 1971), which, despite its title, is largely about the heritage of Aristotelianism.

We have tried to present a consensus view of Aristotle and Aristotelianism. Some modern scholars, however, argue that Aristotle's is a formal logic with no intrinsic connection to the world. See Jan Lukasiewicz, *Aristotle's Syllogistic: From the Standpoint of Modern Formal Logic,* 2d ed. (London: Oxford University Press, 1957).

Chapter Four: Printing and the Rupture of Classification

Until fairly recently, the history of printing has been a scholarly backwater for bibliophiles with little interest in the technology's broader consequences. Two early exceptions are S. H. Steinberg's *Five Hundred Years of Printing* (New York: Criterion Books, 1959) and Lucien Febvre and Henri-Jean Martin's *L'Apparition du livre* (Paris: Albin Michel, 1958), translated as *The Coming of the Book,* ed. Geoffrey Nowell-Smith and David Wooton, trans. David Gerard (London: NLB, 1976). Although these are works of considerable scope and utility, synthesizing a wide range of bibliophilic studies, they nonetheless do not emphasize the cognitive effects of the technology.

By the 1960s, developments in the electronic media began to draw attention to the broader consequences of technological change. One of the earliest and best known explorations of the cognitive effects of printing is Marshall McLuhan's above-cited *Gutenberg Galaxy,* which itself may have been inspired by Walter Ong's magisterial study *Ramus,* also cited above. Ong argues that printing supposedly completes the shift from an oral-aural culture to a visual-spatial one. Mary Carruthers's recent study *The Book of Memory* (Cambridge: Cambridge University Press, 1990) suggests that this oral-visual dichotomy is a bit too pat, obscuring highly visual-spatial aspects of ancient and medieval culture. Instead of using this questionable dichotomy to describe the cognitive effects of typographic literacy, we prefer to emphasize an unduly neglected topic in the history of printing: how the brute abundance of books created information overload.

Elizabeth L. Eisenstein touches on this topic only briefly in her extensive study *The Printing Press as an Agent of Change* (Cambridge: Cambridge University Press, 1979), but lately the topic has begun to draw greater attention. In *The Order of Books,* trans. Lydia G. Cochrane (Stanford: Stanford University Press, 1994), Roger Chartier describes how the surfeit of printed books occasioned a sense of "anxiety." Studies of late-Renaissance encyclopedism also reveal a tension between the traditional desire to order knowledge exhaustively and the practical impossibility of doing so in the wake of printing. In addition to the essays collected in *French Renaissance Studies, 1540–70: Humanism and the Encyclopedia,* ed. Peter Sharratt (Edinburgh: University of Edinburgh Press, 1976), see Neil Kenny, *The Palace of Secrets: Béroalde de Verville and Renaissance Conceptions of Knowledge* (Oxford: Oxford University Press, 1991).

Printing's effect on the literate mind needs to be distinguished from its consequences for the actual spread of literacy, about which relatively little is known for the early modern period. In addition to Harvey J. Graff's above-cited works, *Legacies of Literacy* and *Labyrinths of Literacy,* see his edited collection *Literacy and Social Development in the West: A Reader* (Cambridge: Cambridge University Press, 1981). An excellent local study is David Cressy's *Literacy and the Social Order: Reading and Writing in Tudor and Stuart England* (Cambridge: Cambridge University Press, 1980). Carlo Ginzburg essays the effects of literacy on the mind of a late-sixteenth-century Italian miller in *The Cheese and the Worms,* trans. John and Anne Tedeschi (Baltimore: Johns Hopkins University Press, 1980). R. A. Houston's *Literacy in Early Modern Europe* (London: Longman, 1988) provides a good general survey of early modern literacy.

The information overload that culminates with printing began to build with prior advances in manuscript production. On ancient scroll and codex books, see William Harris's above-cited *Ancient Literacy* and the brief, general study by H. L. Pinner, *The World of Books in Classical Antiquity* (Leiden: A. W. Sijthoff, 1958). The classic study on the growth of medieval document production is M. T. Clanchy, *From Memory to Written Record: England, 1066–1307,* 2d ed. (Oxford: Blackwell, 1993). On medieval techniques for glossing texts see, in addition to the works by Carruthers and Clanchy, Beryl Smalley's *The Study of the Bible in the Middle Ages* (Oxford: Basil Blackwell, 1952). David Knowles, *The Evolution of Medieval Thought* (Baltimore: Helicon Press, 1962) provides a good introduction to scholastic "textbooks" and the tradition of the *summa;* also see the appropriate essays in M.-D. Chenu, O.P., *Nature, Man and Society in the Twelfth Century,* ed. and trans. Jerome Taylor and Lester K. Little (Chicago: University of Chicago Press, 1968). Ivan Illich, *In the Vineyard of the Text* (Chicago: University of Chicago Press, 1993) provides a quick introduction to selected aspects of the culture of the book in the Middle Ages.

Our analysis of classical and Renaissance rhetoric, and the intellectual system of commonplace thought, derives from Zachary Sayre Schiffman, *On the Threshold of Modernity: Relativism in the French Renaissance* (Baltimore: Johns Hopkins University Press, 1991). The best introduction to Renaissance commonplace books is now Ann Moss, *Printed Commonplace Books and the Structuring of Renaissance Thought* (Oxford: Oxford University Press, 1996). Sister Joan Marie Lechner, O.S.U., offers a brief history of the places, from Aristotle through the Renaissance, in her *Renaissance Concepts of the Commonplaces* (New York: Pageant Press, 1962). R. R. Bolgar, *The Classical Heritage and Its Beneficiaries* (Cambridge: Cambridge University Press, 1954) also provides a good introduction to humanist notebook techniques. On reading in antiquity and the Middle Ages, and the ways in which it formed the mind and structured memory, see Carruthers, *Book of Memory,* which also describes the use of *florilegia.* And for an exceptionally cogent analysis of the rhetorical basis of Renaissance humanism, see Hanna H. Gray, "Renaissance Humanism: The Pursuit of Eloquence," *Journal of the History of Ideas* 24 (1963), reprinted in *Renaissance Essays,* ed. Paul O. Kristeller and Philip P. Wiener (New York: Harper & Row, 1968).

Montaigne's response to information overload is detailed in Schiffman, *Threshold of Modernity.* On the role of lists in Rabelais, see Mikhail Bakhtin's extended essay "Forms of Time and the Chronotope in the Novel," in M. M. Bakhtin, *The Dialogic Imagination: Four Essays,* ed. Michael Holquist, trans. Caryl Emerson and Michael Holquist (Austin: University of Texas Press, 1981). Our analysis of Descartes is based on Schiffman, *Threshold of Modernity* and Michael E. Hobart, *Science and Religion in the Thought of Nicolas Malebranche* (Chapel Hill: University of North Carolina Press, 1982). Also see Stephen Gaukroger, *Descartes: An Intellectual Biography* (Oxford: Oxford University Press, 1995). The best guide to early modern mnemonic schemes, including Lambert Schenkel's, remains Frances A. Yates, *The Art of Memory* (Chicago: University of Chicago Press, 1966).

Chapter Five: Numeracy, Analysis, and the Reintegration of Knowledge

The actual document recording Descartes's dream has been lost, but Leibniz had copied several passages from it, having seen it among the papers of one Clerselier, Descartes's literary executor; also, Adrien Baillet, Descartes's biographer, had summarized it in his *Vie de monsieur des Cartes* (1691). Now a classic, the standard historical account of the dream is Jacques Maritain, *The Dream of Descartes,* trans. Mabelle L. Andison (New York: Philosophical Library, 1944). Philip J. Davis and Reuben Hersh in *Descartes' Dream: The World According to Mathematics* (Boston: Houghton Mifflin, 1986) present a contemporary treatment of the Cartesian "nightmare," as they call it.

To date there exists no major study that interprets numeracy and mathematics as information technology in the manner we have adopted. Our interpretation is based on the following types of material: histories of mathematics, early modern science, and philosophy; accounts of modern information technology (noted below); and a broad range of intellectual and cultural studies. Cited in this chapter, and here worth noting as a source offering numerous insights for a historical appreciation of information, is Norbert Wiener, *The Human Use of Human Beings: Cybernetics and Society* (New York: Avon Books, 1967).

As a general introduction to numeracy and mathematical thinking, Tobias Dantzig, *Number: The Language of Science,* 4th ed. (Garden City, N.Y.: Doubleday, 1954) still towers over more recent works with its clarity and interest. Other, excellent surveys of early, premodern counting systems, complete with illustrations and explanations of numerals and ciphers, include Graham Flegg, *Numbers: Their History and Meaning* (London: Andre Deutsch, 1983); John D. Barrow, *Pi in the Sky: Counting, Thinking, and Being* (Oxford: Oxford University Press, 1992); Karl Menninger, *Number Words and Number Symbols: A Cultural History of Numbers,* trans. Paul Broneer (New York: Dover, 1992); and the magisterial work of Florian Cajori, *A History of Mathematical Notations,* 2 vols. (New York: Dover, 1993). In *The Measure of Reality: Quantification and Western Society,*

1250–1600 (Cambridge: Cambridge University Press, 1997), Alfred W. Crosby engagingly depicts the history of "pantometry" in early Europe—the growing penchant for measuring and quantifying—and its influence on the European *mentalité.*

Scholars still debate whether a divide exists between premodern and modern science—the so-called continuity question. In this debate, historians of ancient and medieval science generally claim that in premodern thought and practices there existed genuine "science," and that this science both anticipated much of the modern scientific tradition and contributed substantially to its development. There is thus an evolutionary "continuity" from ancient to medieval to modern science. Opponents of this thesis continue to stress the revolutionary nature of sixteenth- and seventeenth-century developments, especially the creations of a modern worldview and of the experimental and analytical methods associated with modern science. On this reading there was a discontinuity or rupture between medieval and modern science. This interpretation informs our own narrative. H. Floris Cohen's *The Scientific Revolution: An Historiographical Inquiry* (Chicago: University of Chicago Press, 1994) surveys exhaustively the historiography of the entire debate. See, in particular, part 3 for his summary treatment of the "slippery concept" of scientific revolution. See also I. Bernard Cohen's compendious *Revolution in Science* (Cambridge, Mass.: Harvard University Press, 1985), especially sections 1–3.

A good, up-to-date introduction to ancient and medieval science is David C. Lindberg, *The Beginnings of Western Science: The European Scientific Tradition in Philosophical, Religious, and Institutional Context, 600* B.C. *to* A.D. *1450* (Chicago: University of Chicago Press, 1992), whose last chapter offers an intelligent discussion of the continuity debate. Lindberg and Robert S. Westman have also edited a collection of recent articles that rethink the entire question of whether there was indeed a "scientific revolution" in the early modern period: *Reappraisals of the Scientific Revolution* (Cambridge: Cambridge University Press, 1990). Similarly, Peter Dear, ed., *The Scientific Enterprise in Early Modern Europe: Readings from Isis* (Chicago: University of Chicago Press, 1997) has culled from *Isis,* the premier scholarly journal devoted to the history of science, a collection of articles from the past forty-five years representing both traditional and revisionist scholarship on the question. Included there is an essay by Steven Shapin, whose subsequent trim volume *The Scientific Revolution* (Chicago: University of Chicago Press, 1996) richly summarizes current scholarly thinking about the problem. "There was no such thing as the Scientific Revolution," he opens, "and this is a book about it." Other treatments of early modern science on which we have relied are Margaret C. Jacob, *The Cultural Meaning of the Scientific Revolution* (Philadelphia: Temple University Press, 1988), and, for the development of classical "mechanics" and the mechanization of scientific thought, the nonpareil work of E. J. Dijksterhuis, *The Mechanization of the World Picture,* trans. C. Dikshoorn (London: Oxford University Press, 1961).

We mentioned earlier that our interpretation of Descartes rests on our own, previously published scholarship and the sources cited therein. In particular, we have relied heavily on Hobart's above-cited *Science and Religion* for our treatment of the cardinal and ordinal principles of number and their implications for the Cartesian revolution in thought. Stephen Gaukroger's recent, masterful study *Descartes: An Intellectual Biography* (also mentioned earlier) promises to endure as the definitive account of Descartes's intellectual development. Gaukroger's bibliography offers an extensive range of primary and current secondary sources to those with an interest in studying this critical figure further from either a historical or a philosophical perspective.

In our presentation of coordinate geometry, we have relied on the first-rate history of mathematics authored by Edna Kramer, *The Nature and Growth of Modern Mathematics* (Princeton: Princeton University Press, 1980). Kramer not only provides a history of what other mathematicians thought, but also an accessible "how-to" introduction to mathematical thinking for the general reader. Likewise, Jagjit Singh, *Great Ideas of Modern Mathematics: Their Nature and Use* (New York: Dover, 1959) explains with great clarity key mathematical concepts and operations. A useful selection from François Viète's *Introduction to the Analytic Art,* as well as selections from other early figures, together with commentary, may be found in D. J. Struik, ed., *A Source Book in Mathematics: 1200–1800* (Princeton: Princeton University Press, 1986). The standard collection of writings by Descartes is *The Philosophical Works of Descartes,* trans. and ed. Elizabeth S. Haldane and G. R. T. Ross, 2 vols. (Cambridge: Cambridge University Press, 1970). It should be supplemented by Descartes's own geometry, available in translation with a facsimile of the original, as *The Geometry of René Descartes,* trans. David Eugene Smith and Marcia L. Lathan (New York: Dover, 1954). For works of Galileo, see Stillman Drake, ed. and trans., *Discoveries and Opinions of Galileo* (Garden City, N.Y.: Doubleday Anchor Books, 1957), and Galileo Galilei, *Two New Sciences,* trans. Stillman Drake (Madison: University of Wisconsin Press, 1974). Especially useful is Drake's introduction to the latter volume.

Chapter Six: The Analytical World Map

A noteworthy and detailed, though rather Whiggish, account of the development of the calculus is Carl B. Boyer, *The History of the Calculus and Its Conceptual Development* (New York: Dover, 1949). See also the relevant chapters of Morris Kline, *Mathematical Thought from Ancient to Modern Times* (New York: Oxford University Press, 1972) and *Mathematics and the Physical World* (New York: Thomas Y. Crowell, 1959), plus the previously cited works of Kramer, Dantzig, and Singh. These authors all provide accessible introductions to the calculus for those with a modicum of mathematical exposure and a mite of courage for more. Friedrich Waismann, *Introduction to Mathematical Thinking,* trans. Theodore J. Benac (New York: Harper & Row, 1959) is more challenging technically and more rewarding conceptually than the aforementioned.

Though first published in 1937 and somewhat dated, Eric Temple Bell, *Men of Mathematics* (New York: Simon & Schuster, 1986) retains much of its iconoclastic delight. Other popular accounts of modern mathematics and the men (mostly) who developed it include Stuart Hollingdale, *Makers of Mathematics* (New York: Penguin, 1989); Lloyd Motz and Jefferson Hane Weaver, *The Story of Mathematics* (New York: Avon Books, 1993); and John McLeish, *The Story of Numbers: How Mathematics Has Shaped Civilization* (New York: Fawcett Columbine, 1991).

Decades ago, Alfred North Whitehead made the Enlightenment point trenchantly. "Classification," he wrote, "is a halfway house between the immediate concreteness of the individual thing and the complete abstraction of mathematical notions." See his *Science and the Modern World* (New York: New American Library, 1948), 33–34 and passim.

Two other classics on the role of the calculus and mathematics in the Enlightenment repay many readings: Ernst Cassirer, *The Philosophy of the Enlightenment,* trans. Fritz C. A. Koelln and James P. Pettegrove (Boston: Beacon Press, 1955); and Charles Coulston Gillispie's masterful and enduring study *The Edge of Objectivity: An Essay in the History of Scientific Ideas* (Princeton: Princeton University Press, 1960).

Entire libraries are devoted to the general subject of science and the Enlightenment. In a collection of essays, G. S. Rousseau and Roy Porter, eds., *The Ferment of Knowledge: Studies in the Historiography of Eighteenth-Century Science* (Cambridge: Cambridge University Press, 1980) provide a worthwhile introduction. Among works we have found useful are Thomas L. Hankins, *Science and the Enlightenment* (Cambridge: Cambridge University Press, 1985); an earlier article by Robert McCrae, "The Unity of the Sciences: Bacon, Descartes, and Leibniz," *Journal of the History of Ideas* 18 (1957): 27–48; and Keith Michael Baker, *Condorcet: From Natural Philosophy to Social Mathematics* (Chicago: University of Chicago Press, 1975). For various of the figures cited, we have also relied on the following primary and secondary works: François Marie Arouet de Voltaire, *Philosophical Letters,* trans. Ernest Dilworth (Indianapolis: Bobbs-Merrill, 1961); *Leibniz: Selections,* ed. Philip P. Wiener (New York: Charles Scribner's Sons, 1951); C. A. van Peursen, *Leibniz: A Guide to His Philosophy,* trans. Hubert Hoskins (New York: Dutton, 1970); Lester Crocker, *Diderot's Chaotic Order: Approach to Synthesis* (Princeton: Princeton University Press, 1974); Arthur M. Wilson, *Diderot* (New York: Oxford University Press, 1972); Thomas L. Hankins, *Jean d'Alembert: Science and the Enlightenment* (Oxford: Oxford University Press, 1970).

Our interpretation of the *Encyclopedia*'s organization may be found in significantly greater detail in Michael E. Hobart, "The Analytical Vision and Organisation of Knowledge in the *Encyclopédie,*" *Studies on Voltaire and the Eighteenth Century,* no. 327 (1995): 147–75. We are grateful to *Studies on Voltaire and the Eighteenth Century* for permission to draw on this material in the current chapter. For other readings of the "encyclopedic arrangement," see Robert Darnton, "Philosophers Trim the Tree of Knowledge," in *The Great Cat Massacre and Other Episodes in French Cultural History* (New York: Vintage Books, 1985); Cynthia J. Koepp, "The Alphabetical Order: Work in Diderot's *Encyclopédie,*" in Steven Laurence Kaplan and Cynthia J. Koepp, eds., *Work in France: Representations, Meaning, Organization, and Practice* (Ithaca, N.Y.: Cornell University Press, 1986), 229–57; John Lough, *The Encyclopédie* (Geneva: Slatkine Reprints, 1989); Gary I. Brown, "The Evolution of the Term 'Mixed Mathematics'," *Journal of the History of Ideas* 52, no. 1 (1991): 81–102. Beyond the original—Denis Diderot and Jean Le Rond d'Alembert, eds., *Encyclopédie, ou dictionnaire raisonné des sciences, des arts et des métiers, par un société de gens des lettres* (Paris, 1751–65)—d'Alembert's *Preliminary Discourse to the Encyclopedia of Diderot,* trans. Richard N. Schwab (Indianapolis: Bobbs-Merrill, 1963) provides ready access to the editors' analytical vision.

Chapter Seven: Analysis Uprooted

Babbage claimed that the clearest account of the Analytical Engine was provided by his friend and confidante Ada Augusta (1815–52), Byron's daughter and the Countess of Lovelace, in her "Sketch of the Analytical Engine Invented by Charles Babbage." It is included in the useful collection of writings, by and about Babbage, edited by Philip

Morrison and Emily Morrison, *Charles Babbage, On the Principles and Development of the Calculator, and Other Seminal Writings by Charles Babbage and Others* (New York: Dover, 1961). For more detailed biographical information, see Anthony Hyman, *Charles Babbage: Pioneer of the Computer* (Princeton: Princeton University Press, 1983).

The previously cited general histories of mathematics all offer standard introductions to the work of Laplace, Lagrange, and Hamilton (Kramer is exceptionally fine), as well as other nineteenth-century figures. In a model of scientific biography, Thomas L. Hankins, *Sir William Rowan Hamilton* (Baltimore: Johns Hopkins University Press, 1980) brilliantly covers both Hamilton's technical accomplishments and his life and times. As part of a somewhat idiosyncratic account of consciousness, modeled on quantum physics, Roger Penrose also presents an incisive account of Hamilton's work, as well as of other dimensions of classical physics in his broad and sweeping volume *The Emperor's New Mind: Concerning Computers, Minds, and the Laws of Physics* (Oxford: Oxford University Press, 1989).

Nowhere is the confidence and flavor of nineteenth-century science and scientific determinism captured better than by the popular works of the mathematician Henri Poincaré, *Science and Method,* trans. Francis Maitland (New York: Dover, n.d.) and *The Value of Science,* trans. George Bruce Halstead (New York: Dover, 1958). That these two volumes, both written in the early twentieth century, remain in print testifies eloquently to their author's insight, foresight, and sheer brilliance. Poincaré would also pave the way mathematically for our contemporary sciences of complexity, which treat nonlinear and nondetermined dynamical systems and phenomena. See, for example, Ivar Ekeland, *Mathematics and the Unexpected* (Chicago: University of Chicago Press, 1988).

Nineteenth-century mathematics and mathematical logic are cogently interpreted in their historical context in a fine series of articles by Joan L. Richards: "The Art and the Science of British Algebra: A Study in the Perception of Mathematical Truth," *Historia Mathematica* 7 (1980): 343–65; "Augustus De Morgan, the History of Mathematics, and the Foundations of Algebra," *Isis* 78 (1987): 7–30; "Rigor and Clarity: Foundations of Mathematics in France and England, 1800–1840," *Science in Context* 4 (1991): 297–319; "God, Truth, and Mathematics in Nineteenth Century England," in M. J. Nye et al., eds., *The Invention of Physical Science* (Dordrecht: Kluwer, 1992), 51–78. Another, specialized treatment of De Morgan is provided by Helena M. Pycior, "Augustus De Morgan's Algebraic Work: The Three Stages," *Isis* 74 (1983): 211–26.

Besides growth in the calculus and analysis, during the nineteenth century extremely sophisticated and abstract techniques for treating statistical probabilities matured that we have bypassed entirely. A penetrating account of these complementary developments is Ian Hacking, *The Taming of Chance* (Cambridge: Cambridge University Press, 1990).

John Passmore, *A Hundred Years of Philosophy* (Middlesex, England: Penguin, 1970) has excellent chapters on the developments of logic, analysis, and British philosophy in the mid- and late nineteenth century and can serve as a standard introduction to the period. Some study of George Boole, *An Investigation of The Laws of Thought, On which are Founded the Mathematical Theories of Logic and Probabilities* (New York, Dover, n.d.) is essential for appreciating the symbolic turn in logic and language during the nineteenth century, as well as for grasping the rudiments of computer logic in the twentieth. A now classic, and quite technical, account of early developments is C. I. Lewis, *A Survey of Symbolic Logic* (New York: Dover, 1960); see in particular the more accessible chapter 1 for De Morgan, Boole, and their immediate successors. See too William Kneale and Martha Kneale, *The Development of Logic* (Oxford: Oxford University Press, 1962) for a standard and very good history of logic.

Approaching the twentieth century, the volumes of available studies on every topic proliferate in a sort of bibliographic Fibonacci series (where each number in a series is the sum of the previous two numbers: 0,1,1,2,3,5,8,13,21,34, . . .). Among works that we have relied on for our rather cursory coverage of the erosion of analytical foundations are the following: an incisive treatment of logical paradoxes, old and new, by R. M. Sainsbury, *Paradoxes* (Cambridge: Cambridge University Press, 1988), who gives a good account of Russell's paradox and others; W. V. Quine, *The Ways of Paradox and Other Essays*, rev. and enlarged ed. (Cambridge: Harvard University Press, 1976); an accessible explanation of Gödel's theorem by Ernest Nagel and James R. Newman, *Gödel's Proof* (New York: New York University Press, 1958); Douglas Hofstader's fascinating tapestry *Gödel, Escher, Bach: An Eternal Golden Braid* (New York: Vintage Books, 1980); Heinz R. Pagels, *The Cosmic Code: Quantum Physics as the Language of Nature* (New York: Bantam Books, 1982), who has written a lucid historical and conceptual treatment of the revolution in quantum physics and the quanta today; Werner Heisenberg, *Physics and Philosophy: The Revolution in Modern Science* (New York: Harper & Row, 1958).

Chapter Eight: The Realm of Pure Technique

Three authors from three different disciplines—history, science, philosophy—have been critical for helping us situate the power and play of information in the intellectual climate of our age. The first is Johan Huizinga, *Homo Ludens: A Study of the Play Element in Culture* (Boston: Beacon, 1955), whose study remains indispensable for an understanding of play in culture and for many (unintended) insights into our own times. The second is Heinz R. Pagels, *The Dreams of Reason: The Computer and the Rise of the Sciences of Complexity* (New York: Bantam Books, 1989). Although the title suggests but another in the plethora of popular books on contemporary science and computers, this book stands all by itself. As in his other works (including the above-cited book on quantum physics), Pagels explains complicated and technical matters with clarity, insight, and a depth of understanding that never fails to enrich the lay scientist in us. Untimely, as is death's wont, his has deprived us of a graceful, interpretative voice of contemporary scientific culture. The third is Arthur Danto, *Connections to the World: The Basic Concepts of Philosophy* (New York: Harper & Row, 1989), whose critical acumen, erudition, and imaginative prose all contribute to a quality of exposition seldom found among academic philosophers, one appropriate for specialist and general reader alike.

In recent years, histories of computer technology have proliferated nearly apace with the profusion of information itself. In this wealth of material, we have relied heavily on what has become a standard reference: Herman H. Goldstine, *The Computer from Pascal to von Neumann* (Princeton: Princeton University Press, 1972). A colleague of von Neumann at Princeton's Institute for Advanced Studies and a computer pioneer in his own right, Goldstine writes knowingly of developments from the 1930s through the 1960s, as well as about earlier, pre-electronic computing. The latter is well covered in the lengthy chapters and essays of William Aspray, ed., *Computing before Computers* (Ames: Iowa State University Press, 1990) and Joel Shurkin, *Engines of the Mind: A History of the Computer* (New York: W. W. Norton, 1984). Other works chronicling the evolution of the electronic computer from the 1940s to the present include David Ritchie, *The Computer Pioneers: The Making of the Modern Computer* (New York, Simon & Schuster, 1986); René Moreau, *The Computer Comes of Age: The People, the Hardware, and the Software* (Cambridge, Mass.: MIT Press, 1984); Michael R. Williams, *A History of Computing Technology* (Englewood Cliffs, N.J.: Prentice Hall, 1985).

For detailed discussion of computer technology, both analogue and digital, see the fascinating volumes by Francis J. Murray, *Mathematical Machines,* 2 vols. (New York: Columbia University Press, 1961). Volume 1 is titled "Digital Computers" and volume 2 "Analogue Devices." Although quite dated, technical, and somewhat advanced for readers without a bit of background in engineering or mathematics, the volumes retain historical interest because computing technology during the first generation of computers remained within the grasp of the generally educated reader. One might not follow all the technical details, but the "picture" of how computers work was still readily accessible to the imagination. A succinct account of the advantages of electronic computing is provided by John Presper Eckert Jr. and John W. Mauchly in their ENIAC patent application, "Electronic Numerical Integrator and Computer (U.S. Patent)," in John Diebold, ed., *The World of the Computer* (New York: Random House, 1973).

Punditry is a sickness of contemporary culture, and it is not surprising that the proliferation of commentaries about information reflects this broader malaise. A few, however, are noteworthy products of measured reflection, such as classicist and computer specialist J. David Bolter's *Turing's Man: Western Culture in the Computer Age* (Chapel Hill: University of North Carolina Press, 1984). His subsequent effort in *Writing Space: The Computer, Hypertext, and the History of Writing* (Hillsdale, N.J.: Lawrence Erlbaum, 1991) is, however, less informative. Self-styled "philosopher of cyberspace" Michael Heim has written two books, both of which require sympathy for a particular brand of Heidegerrian philosophizing: the aforementioned *Electric Language* and *The Metaphysics of Virtual Reality* (New York: Oxford University Press, 1993). Computer-age enthusiasts include Pamela McCorduck, *The Universal Machine: Confessions of a Technological Optimist* (New York: McGraw-Hill, 1985); Jeremy Campbell, *Grammatical Man: Information, Entropy, Language, and Life* (New York: Simon & Schuster, 1982); and George Gilder, although Gilder taps a deeper and somewhat more reflective vein in *Microcosm: The Quantum Revolution in Economics and Technology* (New York: Simon & Schuster, 1989). For a measured and general, commonsense treatment of information culture, Arnio Penzias, *Ideas and Information: Managing in a High-Tech World* (New York: Simon & Schuster, 1989) stands out in bas-relief against many other available commentaries; so, too, does Robert W. Lucky, *Silicon Dreams: Information, Man, and Machine* (New York: St. Martin's Press, 1989).

For a superb, incisive study of Turing's life and work, we owe a debt to Andrew Hodges, *Alan Turing: The Enigma* (New York: Simon & Schuster, 1983). In a different, but equally rich vein is Steve J. Heims, *John von Neumann and Norbert Wiener: From Mathematics to the Technologies of Life and Death* (Cambridge, Mass.: MIT Press, 1980), who not only traces the achievements of these promethean figures but also depicts their diametrically opposed views on the relation between scientific research and military power. William Poundstone, *Prisoner's Dilemma* (New York: Anchor Books, 1992) discusses his title's problem but also provides abundant detail about von Neumann's life and work and game theory in general. See too a trio of von Neumann's own writings: *The Computer and the Brain* (New Haven: Yale University Press, 1958), *Theory of Self-reproducing Automata,* ed. and completed by Arthur W. Burks (Urbana: University of Illinois Press, 1966), of which the editor's introduction and part 1 are particularly useful, and Arthur W. Burks, Herman H. Goldstine, and John von Neumann, "Preliminary Discussion of the Logical Design of an Electronic Computing Instrument," in the aforementioned Diebold, ed., *World of the Computer.*

Chapter Nine: Information Play

Materials supporting our account of the profusion of play may be found in the general works on computers and information cited above. Additionally, see Nicholas Negroponte, *Being Digital* (New York: Vintage Books, 1996). Tom Forester, ed., *The Information Technology Revolution* (Cambridge, Mass.: MIT Press, 1985) offers a large collection of essays covering virtually every aspect of the computer's influence on contemporary life, and from a range of yay-saying to nay-saying viewpoints. For some fanciful speculations about a future computer-based culture, see in the Forester edition an essay by Yoneji Masuda titled "Computopia." Other futuristic scenarios may be found in Alvin Toffler, *The Third Wave* (New York: Morrow, 1980) and Edward Feigenbaum and Pamela McCorduck, *The Fifth Generation: Artificial Intelligence and Japan's Challenge to the World* (Reading, Mass.: Addison-Wesley, 1983). We have no particular intellectual objections to these and comparable scenarios except that they gloss over the historical dimensions of the contemporary information age—dimensions, to be sure, that are not so easy to discern. A good, down-to-earth alternative to such speculations is Theodore Roszak, *The Cult of Information: A Neo-Luddite Treatise on High-Tech, Artificial Intelligence, and the True Art of Thinking,* 2d ed. (Berkeley: University of California Press, 1994).

Unlike historians' debates surrounding the scientific revolution of the sixteenth and seventeenth centuries, and the "continuity question," commentaries on the new "chaos" in mathematics, the sciences of complexity, and "emergent" phenomena are both less and more strident, and for the same reason. These developments have yet to secure their consensus in the narrative of scientific accomplishment, which one can argue for or against, and their overall claim to significance—whether revolutionary or lacunae filling—is not yet clear. As their proponents might say, the new sciences themselves are emergent phenomena. Thus their detractors tend to pay them scant attention, or, dismissively, none at all; conversely, their proponents at times seem to proselytize innovations with a fervor that would have impressed the likes of Saint Paul.

Many contemporary physicists and other scientists, of course, hold ardently to the reductionist vision spawned in the golden age of classical physics. In Nobel laureate Steven Weinberg they have a worthy and stalwart spokesman. See, for example, the accessible *Dreams of a Final Theory* (New York: Vintage Books, 1994) and *The First Three Minutes: A Modern View of the Origin of the Universe,* updated edition (New York: Basic Books, 1993). Likewise, in *The Character of Physical Law* (Cambridge, Mass.: MIT Press, 1967) and *QED: The Strange Theory of Light and Matter* (Princeton: Princeton University Press, 1985), the late, Nobel-prize-winning physicist Richard Feynman reveals his genius as a teacher and expositor of contemporary physics, as well as one of its foremost theoretical practitioners. John D. Barrow, *Theories of Everything: The Quest for Ultimate*

Explanation (New York: Fawcett Columbine, 1991) gives a lucid summary of the status of contemporary, reductionist efforts to "tie it all together."

Those seeking a general introduction to the new, nonreductionist sciences and, especially, to the mathematics of "chaos" itself are well served by the popular work of James Gleick, *Chaos: Making A New Science* (New York: Penguin, 1987), but even more so by David Ruelle's little gem *Chance and Chaos* (Princeton: Princeton University Press, 1991). Nobel chemist Ilya Prigogine and Isabelle Stengers provide a deeper and broader range of observations and reflections in *Order Out of Chaos: Man's New Dialogue with Nature* (New York: Bantam Books, 1984). Other books explaining chaos theory and its role in accounting for diverse natural phenomena include Ivars Peterson, *Newton's Clock: Chaos in the Solar System* (New York: W. H. Freeman, 1993); Leon Glass and Michael E. Mackey, *From Clocks to Chaos: The Rhythms of Life* (Princeton: Princeton University Press, 1988); Nina Hall, ed., *Exploring Chaos: A Guide to the New Science of Disorder* (New York: W. W. Norton, 1994); David Peak and Michael Frame, *Chaos under Control: The Art and Science of Complexity* (New York: W. H. Freeman, 1994); and Ian Stewart, *Does God Play Dice? The Mathematics of Chaos* (Cambridge: Basil Blackwell, 1989). All the foregoing have been written for a general audience, with little or no background in mathematics required. (Still, some helps.) Of a more technical nature is the related collection of articles in F. R. S. Fleischmann, D. J. Tildesley, and R. C. Ball, eds., *Fractals in the Natural Sciences* (Princeton: Princeton University Press, 1989).

Scientist Peter Coveney and prize-winning science journalist Roger Highfield have combined efforts to author a pair of splendid accounts of complexity and the arrow of time: *Frontiers of Complexity: The Search for Order in a Chaotic World* (New York: Ballantine Books, 1995) and *The Arrow of Time: A Voyage through Science to Solve Time's Greatest Mystery* (New York: Fawcett Columbine, 1991). John L. Casti, *Complexification: Explaining a Paradoxical World through the Science of Surprise* (New York: Harper Collins, 1994) is equally lucid and worthwhile.

Stuart Kauffman lacked a good editor in *At Home in the Universe: The Search for the Laws of Self-organization and Complexity* (New York: Oxford University Press, 1995), but even with clunky prose, the book is as rich, incisive, and fascinating as the subject it presents. Nobel-winning quark discoverer and cofounder of the Sante Fe Institute, Murray Gell-Mann, *The Quark and the Jaguar: Adventures in the Simple and the Complex* (New York: W. H. Freeman, 1994) presents the sciences of complexity in a broad continuum of subjects, ranging from the foundational laws of physics to social and cultural diversity. Other works on complexity and emergence that are well worth the effort include M. Mitchell Waldrop, *Complexity: The Emerging Science at the Edge of Order and Chaos* (New York: Simon & Schuster, 1992); Daniel Stein, ed., *Lectures in the Sciences of Complexity,* Sante Fe Institute Studies in the Sciences of Complexity, Lectures, vol. 1 (Redwood City, Calif.: Addison-Wesley, 1989); Steven Levy, *Artificial Life: How Computers Are Transforming Our Understanding of Evolution and the Future of Life* (New York: Pantheon Books, 1992); Roger Lewin, *Complexity: Life at the Edge of Chaos* (New York: Collier Books, 1992); and Gregoire Nicolis and Ilya Prigogine, *Exploring Complexity* (New York: Freeman, 1989).

Conclusion: The Two Cultures and the Arrow of Time

C. P. Snow ruminates on the divisions of modern intellectual life in his collection of lectures *The Two Cultures and the Scientific Revolution* (New York: Cambridge University Press, 1961). For a good, brief introduction to critical theory and its place in modern philosophy, see Richard Kearney, *Modern Movements in European Philosophy* (Manchester: Manchester University Press, 1986). By now the different schools of thought Kearney delineates—phenomenology, structuralism, and Marxian critical theory—have begun to merge into a general, poststructuralist form of critical theory. One of the foremost voices of this poststructuralist movement is the scholarly journal *Social Text,* whose recently published article by Alan D. Sokal, "Transgressing the Boundaries—Toward a Transformative Hermeneutics of Quantum Gravity," no. 46/47 (1996): 217–52, exemplifies the conflict between the sciences and the humanities. Sokal, a mathematical physicist, wrote the article as a prank—parodying the critical theories that challenge the "objectivity" of science—yet the editors of the journal mistook it as a serious work. For differing views of this incident in the culture wars of the late twentieth century, see the article by the physicist Steven Weinberg, "Sokal's Hoax," *The New York Review of Books* 43, no. 13 (1996): 11–15, along with the subsequent exchange generated by Weinberg's article in *The New York Review of Books* 43, no. 15 (1996): 54–56.

Weinberg shows how critical theorists use the sciences of complexity to undermine the claims of reductive science. For the reverse trend, which subsumes the humanities and social sciences under the new information idiom, see the earlier-cited works: Campbell, *Grammatical Man;* Lewin, *Complexity;* and Kauffman, *At Home in the Universe.* Also see Robert Wright, *Three Scientists and Their Gods: Looking for Meaning in an Age of Information* (New York: Harper & Row, 1988).

John Brockman, *The Third Culture* (New York: Simon & Schuster, 1996) has gathered interviews—many off the internet—from a wide range of contemporary empirical and theoretical scientists, allowing them to present their own work to the public and to comment critically on the research and findings of others. The result is a lively volume devoted to the premise that a "third culture" exists, one that is gradually supplanting C. P. Snow's two cultures of the sciences and the humanities. The third culture, Brockman writes, "consists of those scientists and other thinkers in the empirical world who, through their work and expository writing, are taking the place of the traditional intellectual in rendering visible the deeper meanings of our lives, redefining who and what we are." We shall see.

INDEX

Library of Congress Cataloging-in-Publication Data

Hobart, Michael E., 1944–
Information ages : literacy, numeracy, and the computer revolution / Michael E. Hobart and Zachary S. Schiffman.
p. cm.
Includes bibliographical references and index.
ISBN 0-8018-5881-X (alk. paper)
1. Computers and civilization. 2. Information technology.
I. Schiffman, Zachary Sayre. II. Title
QA76.9C66H63 1998
303.48′34—dc21 98-12764 CIP